Lecture Notes in Mathematics

T0225930

1216

Jacob Kogan

Bifurcation of Extremals in Optimal Control

Springer-Verlag

Berlin Heidelberg New York London Paris Tokyo

Author

Jacob Kogan
Department of Mathematics
The Weizmann Institute of Science
Rehovot, Israel

Mathematics Subject Classification (1980): 49

ISBN 3-540-16818-4 Springer-Verlag Berlin Heidelberg New York
ISBN 0-387-16818-4 Springer-Verlag New York Berlin Heidelberg

Library of Congress Cataloging-in-Publication Data. Kogan, Jacob, 1954- Bifurcation of extremals in optimal control. (Lecture notes in mathematics; 1216) Bibliography: p. Includes index. 1. Control theory. 2. Bifurcation theory. I. Title. II. Series: Lecture notes in mathematics (Springer-Verlag); 1216.
QA3.L28 no. 1216 510 s 86-24878 [QA402.3] [629.8'312]
ISBN 0-387-16818-4 (U.S.)

© Springer-Verlag Berlin Heidelberg 1986
Printed in Germany

Printing and binding: Druckhaus Beltz, Hemsbach/Bergstr.
2146/3140-543210

To my uncle Boris

Preview

The study concerns bifurcations of extremals in optimal control problems. The roots of this topic descends to the classical theory of the calculus of variations. The question of intersection of extremals has been considered in the calculus of variations under a sufficiency criterion, which enables to derive the famous Jacobi necessary condition. This condition guarantees the absence of conjugate points, namely the points of intersection of neighboring extremals.

We show in this work that in optimal control conjugate points exist even under a natural generalization of the sufficiency criterion of the calculus of variations. However, we discover that the set of the conjugate points has a simple and elegant structure.

The set of the conjugate points is described in the study for three different types of optimal control problems: an optimal control system with a scalar cost, an optimal control system with a vector cost functional, and an optimal control problem with constraints. In the case of a linear control equation we find out that the conjugate points are the points where the dimension of the attainable set increases; in particular these points do not depend on a cost functional. In a nonlinear case the conjugate points of an extremal $x(t)$ are determined by the attainable set of the system linearized about the extremal $x(t)$. The Jacobi necessary condition is recovered as a special particular case.

The first chapter of the study is an overview of the concepts, definitions, methods and results, the last, however, without proofs. The remainder of the work contains full proofs of the results.

In conclusion I wish to thank the numerous people for their assistance in this work. I am indebted especially to Zvi Artstein with whom I had many stimulating conversations. My thanks go to good people at Physics Department, University of Toronto for their cooperation and help with TEX typesetting system. Last, but not least I wish to acknowledge the much appreciated support of the ALD Ltd.

Contents

§1.1 The problem

In this study we analyze extremal trajectories of optimal control problems. The purpose of this work is to investigate under what conditions an extremal trajectory branches or bifurcates in time. By an extremal (or an extremal trajectory) of an optimal control problem we mean a trajectory which satisfies a first order necessary condition. In this study we examine trajectories of a control problem that satisfy the Pontryagin Maximum Principle as the necessary condition. We also examine extremals generated by the Euler–Lagrange equations. Connection between extremals generated by the different necessary conditions will be discussed.

The modern optimal control theory has deep roots originating from the classical theory of the calculus of variations. Numerous problems of the calculus of variations were treated under sufficient conditions which guarantee the nonexistence of branching points. In this work we adopt a natural generalization of these conditions and clarify reasons for the absence of branching points in the calculus of variations and, on the other hand, their appearance in problems of optimal control. We discover that when these branching points do occur, they have a nice detectable structure.

The study is organized as follows: This overview takes a general look at the results obtained in the study, indicating links to the fundamental ideas of the classical calculus of variations. The main concepts, ideas and methods are presented in this chapter. The detailed examples here will illustrate what is meant by words like branching and branching points. For the sake of convenience almost all examples are assembled in the last section of the chapter. Only those examples that help to explain new concepts and relations are displayed in the corresponding sections. The remainder of the study is devoted to full proofs of the results along with all the respective details.

§1.2 Background from the calculus of variations

We introduce first, for the sake of convenience, the following convention: In order to distinguish between functions of a real variable and elements of the real Euclidean space R^n, the functions will be denoted throughout by boldface letters, in contrast with points in R^n. Namely x, as well as $x(t)$, denotes a function, and x is an element of R^n. The norm of $x \in R^n$ is denoted by $|x|$, $\langle x, y \rangle$ denotes the scalar product in R^n and $\dot{x}$ denotes differentiation with respect to time, i.e. $\dot{x} = \frac{dx}{dt}$. The norm of $n \times n$ matrix M is denoted by $|M|$.

In the classical theory of the calculus of variations the question of intersection of extremals is considered. The so-called simplest variational problem consists of finding an extremum of a functional of the following form

$$c(x) = \int_{t_1}^{t_2} f(t, x(t), \dot{x}(t)) dt \quad \text{subject to } x(t_1) = x_1, \ x(t_2) = x_2 \tag{1.2.1}$$

where $f(t, x, y)$ is a function with continuous first and second partial derivatives with respect to all its arguments, and the class of admissible curves consists of all smooth curves $x(t)$ joining two points x_1, x_2. There are two significant normed linear spaces which are considered in connection with this variational problem:

1. The space $C[t_1, t_2]$, consisting of all continuous functions $x(t)$ defined on a (closed) interval $[t_1, t_2]$, while the norm is defined as the maximum of the absolute value, i.e.

$$|x|_{C[t_1, t_2]} = \max_{t_1 \le t \le t_2} |x(t)|.$$

2. The space $D_1[t_1, t_2]$, consisting of all functions $x(t)$ defined on an interval $[t_1, t_2]$ which are continuous and have continuous first derivatives. The norm is defined by the formula

$$|x|_{D_1[t_1, t_2]} = \max_{t_1 \le t \le t_2} |x(t)| + \max_{t_1 \le t \le t_2} |\dot{x}(t)|.$$

(We follow here the notions being used by Gelfand and Fomin in [6].) These linear spaces are important in our study. We will emphasize throughout resemblance between the problems of the calculus of variations, which are introduced next, on one hand, and the problems which are considered in our work on the other hand.

A curve $x(t)$ is a weak extremum of (1.2.1) if there exists an $\epsilon > 0$ such that the difference $c(x) - c(y)$ has the same sign for all $y(t)$ in the domain of definition of the functional with $|y - x|_{D_1[t_1, t_2]} < \epsilon$. On the other hand, we say that $x(t)$ is a strong extremum of (1.2.1) if there exists an $\epsilon > 0$ such that $c(x) - c(y)$ has the same sign for all $y(t)$ in the domain of definition of the functional with $|y - x|_{C[t_1, t_2]} < \epsilon$.

An extremal trajectory is usually a good candidate for the extremum. This is the main motivation behind beginning a search for the extremum from investigation the extremal trajectories. In the classical theory of the calculus of variations the integral curves of Euler's equation are called extremals. Namely, $x(t)$ is an extremal of the simplest variational problem if $x(t)$ is a solution of the Euler second order differential equation

$$\frac{d}{dt} f_{\dot{x}}(t, x(t), \dot{x}(t)) = f_x(t, x(t), \dot{x}(t)) \tag{1.2.2}$$

The concept of conjugate point which is recalled next, plays an important role in the derivation of sufficient conditions for a functional to have a weak extremum (see Gelfand and Fomin [6], p. 125, Theorem 6). Consider an extremal curve $x(t)$ defined over the time interval $[t_1, t_2]$. An extremal $y(t)$ is a neighboring extremal if the distance $|y - x|$ is small in the appropriate norm. If the norm is chosen to be $D_1[t_1, t_2]$ then the notion of a conjugate point is the following: The point r is said to be conjugate to the point t_1 if $x(t)$ has a sequence of neighboring extremals drawn from the same initial point $x(t_1)$, such that this sequence tends to $x(t)$, each neighboring extremal intersects $x(t)$ and the points of intersection $(r_i, x(r_i))$ have $(r, x(r))$ as their limit. Namely, there exists a sequence $\{ y_i(t) \}$ of extremals such that

$$\lim_{i \to \infty} |y_i - x|_{D_1[t_1, t_2]} = 0, \; y_i(t_1) = x(t_1), \; y_i(r_i) = x(r_i) \text{ and } \lim_{i \to \infty} (r_i, x(r_i)) = (r, x(r)).$$

The Jacobi necessary condition is the following:

If the extremal $x(t)$ corresponds to a minimum of the functional (1.2.1) and if the matrix $f_{\dot{x}\dot{x}}(t, x(t), \dot{x}(t))$ is positive definite along this extremal, then the open interval (t_1, t_2) contains no points conjugate to t_1. (see [6], p. 124, Theorem 5.)

The theory of the calculus of variations presents also sufficient conditions for the **non–existence** of conjugate points on the interval $[t_1, t_2]$ as follows:

Hypothesis 1.2.1

1. $f_{\dot{x}\dot{x}}(t, x(t), \dot{x}(t))$ is positive definite along the chosen extremal $x(t)$,

2. $\int_{t_1}^{t_2} \begin{pmatrix} h(t) \\ \dot{h}(t) \end{pmatrix}^* \begin{pmatrix} f_{xx}(t, x(t), \dot{x}(t)) & f_{\dot{x}x}(t, x(t), \dot{x}(t)) \\ f_{x\dot{x}}(t, x(t), \dot{x}(t)) & f_{\dot{x}\dot{x}}(t, x(t), x(t)) \end{pmatrix} \begin{pmatrix} h(t) \\ \dot{h}(t) \end{pmatrix} dt$
   is positive definite for each $h(t)$ such that $h(t_1) = h(t_2) = 0$, where * indicates the transpose.

Namely, if the conditions $1 - 2$ hold for the extremal $x(t)$, then the interval $[t_1, t_2]$ contains no point conjugate to t_1 (see [6], p. 123, Theorem 4).

Remark 1.2.1 The nonexistence of conjugate points allows, as it is known from the calculus of variations (see [6], pp. 145–149), the following:

To construct a field of extremals of the functional (1.2.1), namely to derive an ordinary differential equation

$$\dot{x}(t) = \psi(t, x(t)) \tag{1.2.3}$$

such that solutions of (1.2.3) are the extremals of (1.2.1). To define the Hilbert invariant integral. To derive sufficient conditions for a strong extremum. To obtain the Weierstrass necessary conditions for a strong extremum.

This remark emphasizes the importance of the absence of conjugate points in the problems of the calculus of variations. This work is devoted to the investigation of bifurcation of extremals in optimal control. We do not discuss ways to extend the presented in Remark 1.2.1 important constructions. However, in section 1.4 will be shown how to do it in some particular case. In the next section the question of existence of conjugate points in the modern control theory will be discussed.

§1.3 Conjugate points in control theory

Consider an optimal control system

$$\dot{x}(t) = F(t, x(t), u(t)) \tag{1.3.1}$$

where $u : [t_1, t_2] \mapsto R^m$ is measurable and $x : [t_1, t_2] \mapsto R^n$ is absolutely continuous, where measurability is understood to be in the Lebesgue sense, and equalities are always "almost everywhere". Following Berkovitz (see [1], p. 22) we define an admissible pair as follows:

Definition 1.3.1 Let $x(t)$ be an absolutely continuous function from $[t_1, t_2]$ to R^n and $u(t)$ be a measurable function from $[t_1, t_2]$ to R^m. The pair $(x(t), u(t))$ is admissible if it satisfies (1.3.1).

Consider the cost functional defined on the set of admissible pairs $(x(t), u(t))$ as follows:

$$c(x, u) = \int_{t_1}^{t_2} f(t, x(t), u(t)) dt. \tag{1.3.2}$$

We assume throughout that the functions $F(t, x, u)$ and $f(t, x, u)$ have continuous first and second order partial derivatives with respect to (x, u) and measurable in t. That is (for example) $\frac{\partial^2 F}{\partial x_i \partial u_j}(t, x, u)$ is a continuous function in (x, u) for each fixed t, and is a measurable function of t for each fixed (x, u).

One of the problems which is considered in the control theory is the problem of finding an optimum of (1.3.2) among all admissible pairs $(x(t), u(t))$ satisfying the boundary condition $x(t_1) = x_1$, $x(t_2) = x_2$. An extremal trajectory $x(t)$ is a weak optimum of (1.3.2) if there exists a positive scalar ϵ such that $c(x, u) - c(y, w) \leq 0$ for each admissible pair $(y(t), w(t))$ with $x(t_1) = y(t_1)$, $x(t_2) = y(t_2)$ and $|(x, u) - (y, w)| < \epsilon$ in an appropriate norm. In [11], for example, the norm is chosen to be $C[t_1, t_2]$. On the other hand we say that $x(t)$ is a strong optimum of (1.3.2) if there exists a positive ϵ such that $c(x, u) - c(y, w) \leq 0$ for each admissible pair $(y(t), w(t))$ with $x(t_1) = y(t_1)$, $x(t_2) = y(t_2)$ and $|x - y|_{C[t_1, t_2]} < \epsilon$. Hence the problem of finding an optimum of (1.3.2) disintegrates naturally into the problem of finding a weak optimum of (1.3.2), and the problem of finding a strong optimum of (1.3.2).

In order to emphasize the connection with the simplest variational problem we wish to mention, that in the special particular case of the control system $\dot{x}(t) = u(t)$ the two above listed problems coincide with the corresponding problems of the calculus of variations which have been presented in section 1.2.

Sufficient conditions for a weak local optimum in optimal control problems have been discussed, e.g., by Lee and Markus in [11], (see Chapter 5, section 5.2). Sufficient conditions for a strong local optimum in optimal control were derived recently by Zeidan in [22] for Hamiltonians with locally Lipschitz gradients. There has been significant research in this area during the last years (see e.g. [2], [9]). However a connection of an extremal trajectory with conjugate points in optimal problems was not investigated yet. It is shown in this study that extremals in an optimal control problem can intersect each other even under a very natural generalization of Hypothesis 1.2.1 (see Hypothesis 1.3.1). On the other hand, and this is the main contribution of the study, we demonstrate in this work that the branching of extremals in optimal control problems possesses a simple elegant structure. In particular the Jacobi necessary condition is recovered as a special case.

In order to give the formal definition of branching in optimal control problems we need to recall some auxiliary notions.

Definition 1.3.2 A trajectory $x(t)$ is an extremal trajectory of the optimal control problem $(1.3.1) - (1.3.2)$ if $(x(t), u(t))$ is an admissible pair and there exist a scalar $\eta_0 \leq 0$ and an absolutely continuous vector valued function $\eta(t)$ which is defined on $[t_1, t_2]$ and is a solution of the following ordinary differential equation:

$$\frac{d}{dt}\eta(t) = -\eta_0 \frac{\partial f}{\partial x}(t, x(t), u(t)) - \eta(t)\frac{\partial F}{\partial x}(t, x(t), u(t)), \qquad (1.3.3)$$

and such that the triple $(x(t), u(t), \eta(t))$ satisfies the Pontryagin Maximum Principle, i.e. the following additional condition holds on $[t_1, t_2]$:

$$\eta_0 f(t, x(t), u(t)) + \eta(t)F(t, x(t), u(t)) = \max_u \{\eta_0 f(t, x(t), u) + \eta(t)F(t, x(t), u)\}. \qquad (1.3.4)$$

See Berkovitz [1], p. 186.) The last definitions show that an extremal trajectory, which is the main object of the study, usually appears together with an admissible control and a corresponding solution of the adjoint equation (1.3.3). For the sake of the technical convenience we introduce now the notion of an extremal triple as follows:

Definition 1.3.3 A triple $(x(t), u(t), \eta(t))$ is an extremal triple if $(x(t), u(t))$ is an admissible pair, $\eta(t)$ is a solution of (1.3.3) and the triple $(x(t), u(t), \eta(t))$ satisfies condition (1.3.4).

Definition 1.3.4 The Hamiltonian H of the optimal control problem $(1.3.1) - (1.3.2)$ is defined as follows:

$$H(t, x, u, \eta) = \eta_0 f(t, x, u) + \eta F(t, x, u).$$

We present now the definition of a branching point of the extremal trajectory $x(t)$. Suppose that there exists a neighboring extremal $y(t)$ such that $x(t)$ and $y(t)$ coincide over an initial subinterval $[t_1, r]$ of the time interval $[t_1, t_2]$ and differ on the rest of it. Namely, $x(t) = y(t)$ for each $t \in [t_1, r]$ and $x(t) \neq y(t)$ for each $t \in (r, t_2]$. In this case we say that r is a branching point of the extremal trajectory $x(t)$, and $y(t)$ branches out of $x(t)$ at r.

We wish to emphasize that our interest concentrates on the branching points of the extremal trajectory $x(t)$ formed by neighboring extremals initiating at the same point $x(t_1)$, where by a neighboring extremal we mean an extremal $y(t)$ such that the norm $|(x, \eta) - (y, \mu)|_{C[t_1, t_2]}$ is small (how small will be specified later). Here $\eta(t)$, $\mu(t)$ are the corresponding solutions of the adjoint equation (1.3.3). Note, that in the particular case of the control system $\dot{x}(t) = u(t)$ this definition of a neighboring extremal leads to the old definition of a neighboring extremal defined by $D_1[t_1, t_2]$ norm in the calculus of variations.

Let $y(t)$ be an extremal trajectory of the control problem (1.3.1), (1.3.2) with $y(t_1) = x(t_1)$. It is shown in the study (see Chapter 4, section 4.3) that there exists a positive ϵ such that, if $|(x, \eta) - (y, \mu)|_{C[t_1, t_2]} < \epsilon$ and $x(r) \neq y(r)$ for some $r \in [t_1, t_2]$, then $x(t) \neq y(t)$ for each $t \in [r, t_2]$. On the other hand we show by an example (see Example 1.10.1) that, if the condition $|(x, \eta) - (y, \mu)|_{C[t_1, t_2]} < \epsilon$ is not satisfied, the extremal $y(t)$ can be different from $x(t)$ on a subinterval $[r, r + \delta)$ and intersect $x(t)$ once again on $[r + \delta, t_2]$.

In order to present the formal definition of a branching point we need to define rigorously what is meant by a neighboring extremal in this study. We say, first, that an extremal $y(t)$ is an ϵ-neighboring extremal if $|(x, \eta) - (y, \mu)|_{C[t_1, t_2]} < \epsilon$. A point r is a ϵ-branching point of the extremal trajectory $x(t)$ if there exists an ϵ-neighboring extremal $y(t)$ such that

$$x(t) = y(t) \text{ on } [t_1, r], \text{ and } x(t) \neq y(t) \text{ on } (r, t_2].$$

It is clear, that if $\epsilon_1 < \epsilon_2$, then the set of ϵ_1-branching points is a subset of that of ϵ_2-branching points. It is shown in the study (see Chapter 4, section 4.3, Definition 4.3.1), that there exists an $\epsilon^* > 0$ which depends only on the matrix $H_{uu}(t, x(t), u(t), \eta(t))$, such that for each positive δ less than ϵ^*, the set of ϵ^*-branching points coincides with that of δ-branching points. From here on this ϵ^* will define the set of neighboring extremal trajectories as follows:

Definition 1.3.5 An extremal trajectory $y(t)$ is a neighboring extremal if the inequality $|(x, \eta) - (y, \mu)|_{C[t_1, t_2]} < \epsilon^*$ holds.

This definition enables us to define a branching point as follows:

Definition 1.3.6 A point r is a branching point of the extremal trajectory $x(t)$ if there exists a neighboring extremal trajectory $y(t)$ such that

$$x(t) = y(t) \text{ on } [t_1, r], \text{ and } x(t) \neq y(t) \text{ on } (r, t_2].$$

Our main purpose in this work is: to determine the conditions under which the branching points of an extremal $x(t)$ exist or do not exist and to characterize the set of branching points of

$x(t)$. We hope that the knowledge of the branching points will be useful for further development of the modern control theory. By using this knowledge we will show in section 1.5, that in some particular case, namely when the set of branching point is the empty set, all the constructions which have been described in Remark 1.2.1 can be transferred from the calculus of variations to the control theory.

In order to achieve the proposed goal we adopt a natural generalization of the sufficient conditions of the calculus of variations, namely we assume that the following holds for the extremal triple $(x(t), u(t), \eta(t))$.

Hypothesis 1.3.1

1. There exists a positive scalar m such that for each vector $v \in R^m$

$$v^* H_{uu}(t, x(t), u(t), \eta(t))v \leq -m|v|^2 \text{ along } x(t),$$

and the matrix $H_{uu}(t, x(t), u(t), \eta(t))$ is uniformly continuous with respect to t. Namely, for each positive ρ there exists a positive δ such that

$$|H_{uu}(t, x(t), u(t), \eta(t)) - H_{uu}(t, y, w, \xi)| < \rho$$

provided

$$|(t, x(t), u(t), \eta(t)) - (t, y, w, \xi)| < \delta.$$

2. $\displaystyle\int_{t_1}^{t_2} \begin{pmatrix} z(t) \\ v(t) \end{pmatrix}^* \begin{pmatrix} H_{xx}(t, x(t), u(t), \eta(t)) & H_{ux}(t, x(t), u(t), \eta(t)) \\ H_{xu}(t, x(t), u(t), \eta(t)) & H_{uu}(t, x(t), u(t), \eta(t)) \end{pmatrix} \begin{pmatrix} z(t) \\ v(t) \end{pmatrix} dt$

is negative definite for each $v(t)$, $z(t)$ such that

$$z(t) = \int_{t_1}^{t} \Phi(t, \sigma)B(\sigma)v(\sigma)d\sigma, \; z(t_2) = 0,$$

where $A(t)$, $B(t)$ are Lebesgue integrable matrices given by $A(t) = \frac{\partial F}{\partial x}(t, x(t), u(t))$, $B(t) = \frac{\partial F}{\partial u}(t, x(t), u(t))$, and $\Phi(t, t_1)$ is the transition matrix of $\frac{d}{dt}\phi = A(t)\phi$, i.e. $\phi(t) = \Phi(t, t_1)x_1$ is the solution of the equation $\frac{d}{dt}\phi = A(t)\phi$ with $\phi(t_1) = x_1$.

In other words: $z(t)$ is a solution of the linear equation

$$\dot{z}(t) = A(t)z(t) + B(t)v(t)$$

with boundary conditions $z(t_1) = z(t_2) = 0$.

We wish to clarify now the connection between Hypothesis 1.3.1 and Hypothesis 1.2.1 of the calculus of variation. To this end consider the special case of the control system $\dot{x}(t) = u(t)$ and the cost function $f(t, x, u)$ with continuous first and second partial derivatives with respect to all its arguments. We will show, that in this particular case Hypothesis 1.3.1 is equivalent to

Hypothesis 1.2.1. Namely, an extremal trajectory $x(t)$ satisfies Hypothesis 1.3.1 if and only if it satisfies Hypothesis 1.2.1.

Note, first, that in the case of the control system $\dot{x}(t) = u(t)$ the second conditions of the hypotheses are identical. Let $x(t)$ be an extremal which satisfies Hypothesis 1.3.1. The first condition of Hypothesis 1.3.1 implies that $f_{\dot{x}\dot{x}}(t, x(t), \dot{x}(t))$ is positive definite along $x(t)$. Hence, the extremal $x(t)$ satisfies Hypothesis 1.2.1. On the other hand, assume that an extremal $x(t)$ satisfies Hypothesis 1.2.1. Due to continuity properties of $f_{\dot{x}\dot{x}}(t, x, y)$ with respect to all its arguments, one can easily demonstrate that the first condition of Hypothesis 1.3.1 is fulfilled.

In the remainder of the chapter we display results of the study and demonstrate a connection between branching points in the control theory and conjugate points in the calculus of variations. The reasons for existence of branching points in control problems in contrast with their absence in the calculus of variations will become clear.

Our first result concerns a linear control system with a convex cost functional. On the other hand, as we will show later, in the case of a nonlinear control system the branching points of the extremal trajectory $x(t)$ are determined by the linearized system about the extremal $x(t)$. Hence, characterization of the branching points in this simplest case is a significant step forward.

§1.4 The linear control equation

The first result of the study concerns the linear control system

$$\dot{x}(t) = A(t)x(t) + B(t)u(t) \tag{1.4.1}$$

with the cost functional (1.3.2), where in addition $f(t, x, u)$ is assumed to be a strictly convex function in (x, u). The convexity of $f(t, x, u)$ ensures the following: if $x(t)$, $y(t)$ are extremals such that

$$x(t_1) = y(t_1) \text{ and } x(s) = y(s) \text{ for some } s \in [t_1, t_2]$$

then $u(t) = w(t)$ and $x(t) = y(t)$ on $[t_1, s]$. (This statement is proved in the Appendix.) This implies, in particular, that branching is indeed a unique possible form of intersection of extremals initiating at the same point x_1. In the next section we demonstrate that the last statement is also true in the calculus of variations. In the end of section 1.5 we shall reveal the relationship between the notions of conjugate point and branching point.

There is an additional important property of a linear equation and a convex cost which will be shown in the study: Branching points of an extremal trajectory $x(t)$ do not depend on a particular choice of a family of neighboring extremals, i.e. particular choice of an ϵ. Namely, r is a branching point of $x(t)$ if there exists any extremal $y(t)$ which branches out of $x(t)$ at r.

In order to illustrate the phenomenon of branching we introduce an example. In spite of its simplicity, this example, as will be shown in this section, presents a typical case of branching.

Example 1.4.1 A linear control system with a non empty set of branching points.

Consider the linear control system:

$$\frac{d}{dt}x(t) = B(t)u(t)$$

with the cost functional

$$c(x, u) = \int_0^1 u^2(t)dt$$

with $x \in R^2$, u scalar and where $B(t)$ is given by:

$$\begin{pmatrix} 1 \\ 0 \end{pmatrix} \text{ for } 0 \leq t < \tfrac{1}{3}, \quad \begin{pmatrix} 0 \\ 0 \end{pmatrix} \text{ for } \tfrac{1}{3} \leq t < \tfrac{2}{3}, \quad \begin{pmatrix} 0 \\ 1 \end{pmatrix} \text{ for } \tfrac{2}{3} \leq t \leq 1.$$

An extremal control in this case has the following form:

$$u(t) = B^*(t)\eta^*(t)$$

where $\eta(t)$ is a solution of the adjoint ordinary differential equation $\dot{\eta}(t) = 0$ (see Lee and Markus [11], p. 180) and * indicates the transpose. Note, that $\eta(t) = \eta(0)$ and we shall denote it simply by η. It then follows that $u(t) = B(t)^*\eta^*$ for each $t \in [0, 1]$. An extremal trajectory $x(t)$ is, therefore, given by:

$$x(t) = x(0) + \int_0^t B(s)B^*(s)\eta^* ds.$$

The product $B(t)B^*(t)$ can easily be computed and the derived extremal trajectory $x(t)$ has the following form:

$$x(t) = x(0) + \gamma(t)$$

where the vector $\gamma(t)$ is given by:

$$\begin{pmatrix} t\eta_1 \\ 0 \end{pmatrix} \text{ for } 0 \leq t < \tfrac{1}{3}, \quad \begin{pmatrix} \tfrac{1}{3}\eta_1 \\ 0 \end{pmatrix} \text{ for } \tfrac{1}{3} \leq t < \tfrac{2}{3}, \quad \begin{pmatrix} \tfrac{1}{3}\eta_1 \\ (t - \tfrac{2}{3})\eta_2 \end{pmatrix} \text{ for } \tfrac{2}{3} \leq t \leq 1.$$

Here η_1, η_2 are the components of the vector η.

We would like to study the extremal trajectories initiating at the same initial point $x(0)$. The definition of $\gamma(t)$ implies that for each pair of different vectors η, μ such that $\eta_1 = \mu_1$ and $\eta_2 \neq \mu_2$, the corresponding extremal trajectories $x(t)$ and $y(t)$ coincide on the time interval $[0, \tfrac{2}{3}]$ and are different on $(\tfrac{2}{3}, 1]$. Namely, if, for instance, $x(0) = 0$, $\eta = (1, 0)$, and $\mu = (1, 1)$ then $x(t)$ is given by:

$$\begin{pmatrix} t \\ 0 \end{pmatrix} \text{ for } 0 \leq t < \tfrac{1}{3}, \quad \begin{pmatrix} \tfrac{1}{3} \\ 0 \end{pmatrix} \text{ for } \tfrac{1}{3} \leq t < \tfrac{2}{3}, \quad \begin{pmatrix} \tfrac{1}{3} \\ 0 \end{pmatrix} \text{ for } \tfrac{2}{3} \leq t \leq 1,$$

and $y(t)$ is given by:

$$\begin{pmatrix} t \\ 0 \end{pmatrix} \text{ for } 0 \leq t < \tfrac{1}{3}, \quad \begin{pmatrix} \tfrac{1}{3} \\ 0 \end{pmatrix} \text{ for } \tfrac{1}{3} \leq t < \tfrac{2}{3}, \quad \begin{pmatrix} \tfrac{1}{3} \\ t - \tfrac{2}{3} \end{pmatrix} \text{ for } \tfrac{2}{3} \leq t \leq 1,$$

i.e. $x(t) = y(t)$ on $[0, \tfrac{2}{3}]$, and $x(t) \neq y(t)$ on $(\tfrac{2}{3}, 1]$.

Suppose that different extremal trajectories $x(t)$, $y(t)$ coincide on a certain subinterval $[0, \tau]$ of $[0, 1]$. We denote by η and μ the corresponding solutions of the adjoint differential equation. In this case $\eta_1 = \mu_1$. On the other hand, inasmuch as $x(t)$ and $y(t)$ are different, $\eta_2 \neq \mu_2$. This means that $\tau = \tfrac{2}{3}$.

The above considerations prove that $\tfrac{2}{3}$ is the unique branching point of the considered optimal control problem. Hence, the set of the branching points of this control problem is a finite set. As it will be shown in the next section, this result is not incidental, but rather holds for a general linear control system with a convex cost. Namely, the set of branching point is a finite set which does not depend on a chosen extremal.

Derivation of the matching condition.

In the remainder of the section we show that the set of branching points of an extremal trajectory $x(t)$ is a finite set. In order to achieve this goal we need to develop the following auxiliary chain of formulas and definitions. Note, that in this linear case η_0 which enters the Definition 1.3.2, Definition 1.3.3 and Definition 1.3.4 is always negative (see Lee and Markus, [11], p. 181). Only the ratio $\frac{\eta}{\eta_0}$ enters these definitions and so, from now on, we set $\eta_0 = -1$ for convenience.

Let $(y(t), w(t), \mu(t))$ be an extremal triple such that an extremal trajectory $y(t)$ coincides with $x(t)$ on $[t_1, \tau]$, namely the following relation holds:

$$x(t) = y(t) \text{ on } [t_1, \tau]. \tag{1.4.2}$$

Consider the corresponding to the extremals $x(t)$ and $y(t)$ solutions $\eta(t)$, $\mu(t)$ of the adjoint equation (1.3.3), i.e.

$$\dot{\eta}(t) = \frac{\partial f}{\partial x}(t, x(t), u(t)) - \eta(t)A(t),$$

$$\dot{\mu}(t) = \frac{\partial f}{\partial x}(t, y(t), w(t)) - \mu(t)A(t).$$

The equation (1.4.2) implies in this linear case that $u(t) = w(t)$. Hence

$$\frac{d}{dt}(\mu(t) - \eta(t)) = (\mu(t) - \eta(t))A(t) \text{ on } [t_1, \tau].$$

Namely,

$$\mu(t) - \eta(t) = (\mu(t_1) - \eta(t_1))\Phi(t_1, t) \text{ on } [t_1, \tau], \tag{1.4.3}$$

where $\Phi(t, t_1)$ denotes the fundamental solution of the linear differential equation $\dot{x}(t) = A(t)x(t)$ with $\Phi(t_1, t_1) = I$, and I denotes the identity matrix. On the other hand the equation (1.3.4) yields the following relation on $[t_1, r]$:

$$(\mu(t) - \eta(t))B(t) = 0,$$

or, by using formula (1.4.3),

$$(\mu(t_1) - \eta(t_1))\Phi(t_1, t)B(t) = 0 \text{ on } [t_1, r]. \tag{1.4.4}$$

In order to continue the derivation we need now to give the formal definitions of the space of admissible controls of the linear control system (1.4.1).

Definition 1.4.1 The space of all admissible controls $v(t)$ which are defined on the interval $[t_1, r]$ will be denoted by $U[t_1, r]$. Namely, $U[t_1, r]$ consists of all measurable functions $v(t)$ defined on the interval $[t_1, r]$ with range in R^m such, that there exists an absolutely continuous function $\phi(t)$ defined on $[t_1, r]$ with range in R^n which is a solution of the differential equation (1.4.1), i.e.

$$\dot{\phi}(t) = A(t)\phi(t) + B(t)v(t) \text{ on } [t_1, r].$$

We wish to mention, that in accordance with Definition 1.3.1, the pair $(\phi(t), v(t))$ is an admissible pair of the linear system (1.4.1). From this point of view Definition 1.4.1 is a particular case of Definition 1.3.1. The equation (1.4.4) and Definition 1.4.1 imply the following relation:

$$(\mu(t_1) - \eta(t_1))\Phi(t_1, t)B(t)v(t) = 0 \text{ on } [t_1, r] \text{ for each } v \in U[t_1, r].$$

Hence

$$(\mu(t_1) - \eta(t_1)) \int_{t_1}^{r} \Phi(t_1, t)B(t)v(t)dt = 0, \quad \text{for every } v \in U[t_1, r]. \tag{1.4.5}$$

This formula (1.4.5) is the key equation for the description of the branching points. Namely, as we will show in the study (see Chapter 2, section 2.3, Lemma 2.3.1), the extremals $x(t)$ and $y(t)$ coincide on the interval $[t_1, r]$ if and only if the relation (1.4.5) holds. In view of these facts, we call (1.4.5) the matching condition.

<u>Necessary and sufficient conditions for branching.</u>

In order to derive the necessary and sufficient conditions for branching introduction of some additional auxiliary definitions is needed. First, for each subset E of the n dimensional real space R^n we denote by $Sp\{E\}$ the linear subspace of R^n spanned by E, i.e.

$$Sp\{E\} = \{ x \mid x = \sum_{i=1}^{k} \alpha_i e_i, \ e_i \in E, \ \alpha_i \text{ is a scalar}, \ k = 1, 2, \dots \}$$

For each subspace V of R^n we denote by $V^\perp$ its orthogonal complement in R^n. We introduce now the family of vector spaces $V(t_1, s)$ which determines branching points of extremal trajectories.

Definition 1.4.2 For each real number s we define two vector spaces $V(t_1, s)$ and $V(t_1, s^+)$ by the following formulas:

$$V(t_1, s) = Sp\{ \int_{t_1}^{s} \Phi(t_1, t)B(t)v(t)dt \mid v \in U[t_1, s] \},$$

$$V(t_1, s^+) = \bigcap_{h>0} V(t_1, s+h).$$

Denote the dimension of $V(t_1, s)$ by $dimV(t_1, s)$. Consider $dimV(t_1, s)$ as a function of s. Choose s_1, s_2 such that $t_1 \le s_1 \le s_2$ holds. Definition 1.4.2 implies that $V(t_1, s_1)$ is a subspace of $V(t_1, s_2)$. On the other hand for each s the space $V(t_1, s)$ is a subspace of R^n and, therefore, $dimV(t_1, s) \le n$. The above consideration shows that $dimV(t_1, s)$ is a monotone non decreasing step function with respect to s. This implies that the number of its discontinuity points is finite. Note, that $V(t_1, s) \subset V(t_1, s^+)$; but $V(t_1, s) \ne V(t_1, s^+)$ if and only if s is a point of discontinuity of $dimV(t_1, s)$. The points of discontinuity of $dimV(t_1, s)$ which we denote by

$$t_1 < r_1 < ... < r_k < t_2 \tag{1.4.6}$$

play an extremely important role in the description of the set of branching points. In the next statement the importance of the set (1.4.6) will become clear.

The equation (1.4.5) and Definition 1.4.2 yield:

Lemma 1.4.1 If the extremal trajectories $x(t)$ and $y(t)$ coincide on the initial subinterval $[t_1, r]$, then $\mu(t_1) - \eta(t_1) \in V(t_1, r)^\perp$.

It is shown in the study (see Chapter 2, section 2.3, Lemma 2.3.1) that the condition

$$\mu(t_1) - \eta(t_1) \in V(t_1, r)^\perp$$

is the **necessary and sufficient condition** for the matching of the extremals $x(t)$ and $y(t)$ on the initial subinterval $[t_1, r]$.

The main result in the linear case is as follows:

Theorem 1.4.1 The point r is a branching point of the extremal trajectory $x(t)$ if and only if $V(t_1, r)$ is a proper subspace of $V(t_1, r^+)$.

This theorem immediately yields the following corollary:

Corollary 1.4.1 The set of the branching points of the linear optimal control problem (1.4.1) − (1.3.2) depends on the linear control system only (namely the matrices $A(t)$ and $B(t)$). It does not depend on the cost functional or the chosen extremal $x(t)$. The number of branching points does not exceed the dimension of the state space.

This result is, to my opinion, somewhat unexpected: indeed, an extremal trajectory depends on the control system and the cost functional, the extremal trajectories form the branching points, but the branching points depend on the linear control equation only.

Reasons for the existence of branching points.

Our purpose now is to clarify the reasons for the existence of branching points. The structure of the branching of $x(t)$ is closely related to the dimension of the attainable set which is recalled next. The attainable set at time t for the control system with initial time t_1 and initial state x_1 is the set of all points $x \in R^n$ such that for some trajectory $\varphi(s)$ satisfying the condition $\varphi(t_1) = x_1$, the relation $\varphi(t) = x$ holds. In the case when the initial state x_1 is chosen to be 0 the attainable set at time t is a linear subspace of R^n which is denoted by $AT(t_1,t)$.

The dimension of the attainable set is a non decreasing bounded step function of t. This function has, therefore, a finite set of discontinuity points. It is shown in the present work that this set is exactly the finite partition (1.4.6).

At this point we wish to mention that

$$AT(t_1,s) = Sp\{\int_{t_1}^{s} \Phi(s,t)B(t)v(t)dt, \ v \in U[t_1,s]\}.$$

Hence, there is an evident relation between the sets $V(t_1,s)$ and $AT(t_1,s)$, namely

$$V(t_1,s) = \Phi(t_1,s)AT(t_1,s). \tag{1.4.7}$$

Since $\Phi(t_1,s)$ is nonsingular, it follows that $dimV(t_1,s) = dimAT(t_1,s)$. This implies that the partition (1.4.6) is indeed the set of discontinuity points of the function $dimAT(t_1,s)$, as a function of s. Then Theorem 1.4.1 and relation (1.4.7) yield the additional corollary:

Corollary 1.4.2 A point r is a branching point of the extremal trajectory $x(t)$ if and only if r is a point of discontinuity of the function $dimAT(t_1,s)$ as a function of s.

In order to understand the phenomenon of branching we wish to quote the part of Definition 1.3.2: "a trajectory $x(t)$ is an extremal trajectory if there exists an absolutely continuous function $\eta(t)$ which is a solution of equation (1.3.3) such that condition (1.3.4) holds." Namely, the Pontryagin Maximum Principle guarantees existence of a solution of the adjoint equation, but does not answer the question "How many such solutions do exist?" We will, now, answer this question.

The geometric meaning of $\eta(s)$ is as follows: the vector $(-1,\eta(s))$ is a support vector to the extended attainable set at $(\int_{t_1}^{s} f(t,x(t),u(t))dt, x(s))$, where by extended attainable set we mean, as it is customary (see e.g., [11], p. 171), the set of all pairs (c,z) such that c is a scalar, $z \in R^n$ and there exists an admissible pair $(y(t), w(t))$ satisfying

$$y(t_1) = x_1, \ y(s) = z \text{ and } c = \int_{t_1}^{s} f(t,y(t),w(t))dt.$$

It is clear that if $dim AT(t_1, s) = n$, then there exists a unique support vector of the form $(-1, \eta(s))$ to the extended attainable set at $(\int_{t_1}^{s} f(t, x(t), u(t))dt, x(s))$. But if $dim AT(t_1, s) < n$ the situation is different, i.e., there exists a set of support vectors which we denote by $\{ (-1, \eta(s)+v) \}$. Note, that the vectors v form a linear space and its dimension is $n - dim AT(t_1, s)$, (it is shown in Chapter 8, section 8.3). Let $\Xi(s)$ be the set of the vectors $\xi \in R^n$ such that each solution $\mu(t; \xi)$ of the adjoint equation (1.3.3) with $\mu(t_1; \xi) = \xi$ together with the extremal pair $(x(t), u(t))$ satisfies the Pontryagin condition (1.3.4) on the initial interval $[t_1, s]$. Namely, the triple $(x(t), u(t), \mu(t; \xi))$ is an extremal triple of the control problem (1.4.1), (1.3.2) on $[t_1, s]$. It is shown in this work, (see Chapter 8, section 8.3), that the vector $(-1, \eta(s) + v)$ is a support vector to the extended attainable set at $(\int_{t_1}^{s} f(t, x(t), u(t))dt, x(s))$ if and only if

$$v\Phi(s, t_1) \in \{ \Xi(s) - \eta(t_1) \},$$

where $\{ \Xi(s) - \eta(t_1) \}$ denotes the set of all vectors $\{ \xi - \eta(t_1) \mid \xi \in \Xi(s) \}$. This implies that $\{ \Xi(s) - \eta(t_1) \}$ is a linear subspace of R^n and

$$dim\{ \Xi(s) - \eta(t_1) \} = n - dim AT(t_1, s) = dim V(t_1, s)^{\perp}.$$

This relation provides an insight into the connection between the set of the solutions of the adjoint equation corresponding to the extremal pair $(x(t), u(t))$ and the attainable set at time s. Namely, if the attainable set "blows up" at a point s then the set of solutions $\Xi(s)$ "subsides" at s, and a set of extremals branches out of $x(t)$ at s. In particular, the set of branching points of the extremal $x(t)$ is the empty set if and only if the dimension of the attainable set is a constant on (t_1, t_2).

The last statement presents "almost" generalization of the Jacobi necessary condition which appears in the calculus of variations. Namely, the attainable set of the linear control system $\dot{x}(t) = u(t)$, which is actually considered in the calculus of variations, is the state space R^n at each $t \in (t_1, t_2)$. Hence its dimension is, indeed, a constant on (t_1, t_2) and then, due to Corollary 1.4.2, if $f(t, x, u)$ is convex in (x, u), the branching points of this system is the empty set. We wish to emphasize, that in the calculus of variations the function $f(t, x, \dot{x})$ is not necessary a convex function in $(x, \dot{x})$. For this reason the obtained result is only "almost" generalization of the Jacobi condition. The "true" generalization will be presented in the next section.

§1.5 The nonlinear system

In order to study branching points of an extremal trajectory $x(t)$ in the case of the nonlinear control problem (1.3.1) − (1.3.2) we wish to use the Pontryagin Maximum Principle and the technique of representation of the extremals as the solutions of the Hamiltonian system (1.5.1), (1.5.2) which is displayed below. Such a consideration is applicable to a wide class of optimal control problems where $f(t, x, u)$ grows faster then $F(t, x, u)$ as function of u.

Framework.

In order to realize the announced plan the following assumption is introduced:

Assumption 1.5.1 For each triple (t, x, η) close to $(t, x(t), \eta(t))$ there exists a unique $u(t, x, \eta)$ continuously differentiable with respect to (x, η) and measurable in t such that

$$-f(t, x, u(t, x, \eta)) + \eta F(t, x, u(t, x, \eta)) = \max_u \{-f(t, x, u) + \eta F(t, x, u)\} \text{ on } [t_1, t_2].$$

In particular bilinear control systems with quadratic costs, and control problems where the control enters linearly in the dynamics and such that the cost function $f(t, x, u)$ satisfy the condition:

$$f(t, x, u) \geq \alpha |u|^p \text{ for } \alpha > 0, \ p > 1$$

present examples of control problems which fulfill the Assumption 1.5.1.

We follow here the approach which has been used by Kalman in [10]. This approach enables one to reduce the study of extremals initiating at the same point x_1 to the study of the solutions $\{x(t; \xi), \ \eta(t; \xi)\}$ of the following Hamiltonian system:

$$\dot{x}(t; \xi) = H_\eta \tag{1.5.1}$$

$$\dot{\eta}(t; \xi) = -H_x. \tag{1.5.2}$$

Here $x(t_1; \xi) = x_1, \eta(t_1; \xi) = \xi$, and by "reducing" we mean the following: A triple $(x(t), u(t), \eta(t))$ is an extremal triple of the optimal control problem (1.3.1), (1.3.2) if and only if

$$\dot{x}(t) = H_\eta(t, x(t), u(t), \eta(t))$$

$$\dot{\eta}(t) = -H_x(t, x(t), u(t), \eta(t)).$$

We wish to emphasize the connection between the Hamiltonian system and the system of differential equations (1.3.1), (1.3.3). Due to Assumption 1.5.1 there exists a function $u(t, x, \eta)$ defined in a neighborhood of the curve $(t, x(t), \eta(t))$ such that the equation enters Assumption 1.5.1 holds. The substitution of this $u(t, x, \eta)$ into the system (1.3.1), (1.3.3) transforms it to the Hamiltonian system (1.5.1), (1.5.2). This representation of the extremals as solutions of the Hamiltonian system will be extremely useful in our study. In the next subsection we present the reasons for existence of the branching points.

Structure of the set of branching points.

Select some ξ_1 and consider an extremal trajectory $x(t; \xi_1)$. In the case of the linear control equation, the collection of all such $\xi \in R^n$, that the generated extremal $x(t; \xi)$ coincide with $x(t; \xi_1)$ on $[t_1, s]$ was denoted by $\Xi(s)$. As it was mentioned in the preceding section, this set $\Xi(s)$ has the global linear structure for each $s \in [t_1, t_2]$.

We wish now to extend this definition of $\Xi(s)$ for the nonlinear case. As it was mentioned in section 1.3, we are interested in the branching points of the extremal trajectory $x(t; \xi_1)$ formed by the neighboring extremals. This fact encourages us to denote by $\Xi(s)$ the set of all such vectors $\xi \in R^n$, which generates a neighboring extremal trajectory $x(t; \xi)$ with $x(t; \xi_1) = x(t; \xi)$ on $[t_1, s]$.

We wish to remind the reader, that in the linear case the set of the neighboring extremals coincides with the set of the extremal trajectories. Hence, the new definition of $\Xi(s)$ is consistent with the old one.

It is clear that $s \le \sigma$ implies that $\Xi(s) \supset \Xi(\sigma)$. We show (see Chapter 4, section 4.3, Lemma 4.3.1) that for each $s \in [t_1, t_2]$ the set $\Xi(s)$ has a linear structure, and that there exists a finite partition

$$t_1 = r_0 < r_1 < \ldots < r_k < r_{k+1} = t_2 \tag{1.5.3}$$

of the time interval $[t_1, t_2]$ such that for each $s, \sigma \in [t_1, t_2]$ the following two conditions hold:

1. if $s, \sigma \in (r_i, r_{i+1}]$ then $\Xi(s) = \Xi(\sigma)$,

2. if $s \in (r_i, r_{i+1}]$, $\sigma \in (r_j, r_{j+1}]$, $i < j$ then $\Xi(s) \supset \Xi(\sigma)$ and $\Xi(s) \ne \Xi(\sigma)$.

(See Chapter 4, section 4.3, Theorem 4.3.3.)

This implies that the set $\{r_i\}_{i=1}^k$ is the set of the branching points of the extremal trajectory $x(t; \xi_1)$. On the other hand, the structure of the branching of $x(t; \xi_1)$ is closely related to the dimension of the the attainable set of the linearized system about the extremal $x(t; \xi_1)$. Namely, the attainable set of the linear control system

$$\dot{z}(t) = A(t)z(t) + B(t)v(t), \tag{1.5.4}$$

where

$$A(t) = \frac{\partial F}{\partial x}(t, x(t; \xi_1), u(t, x(t; \xi_1), \eta(t, \xi_1))), \quad B(t) = \frac{\partial F}{\partial u}(t, x(t; \xi_1), u(t, x(t; \xi_1), \eta(t, \xi_1))).$$

The following theorem describes the branching points of the extremal trajectory $x(t)$:

Theorem 1.5.1 If the Hypothesis 1.3.1 holds for the chosen extremal trajectory $x(t)$, then the set of the branching points of $x(t)$ coincides with that of the linear control system (1.5.4).

In the remainder of the section the connection between branching points in the control theory and the conjugate points in the calculus of variations will be discussed. We wish to remind the reader that the Jacobi condition describes conjugate points of extremals in the simplest variational problem; namely conjugate points of solutions of the Euler equation. On the other hand, Theorem 1.5.1 deals with solutions of the Pontryagin equation. In the next subsection we demonstrate the connection between the sets of extremals generated by the Euler equation on the one hand, and the Pontryagin equation on the other hand.

Relationship with the Euler extremals.

We introduce now the notion of an Euler extremal in an optimal control problem as follows: a trajectory $x(t)$ is an Euler extremal of the optimal control system (1.3.1) with the cost (1.3.2) if, instead of the condition (1.3.4), the following condition satisfied:

$$H_u(t, x(t), u(t), \eta(t)) = 0 \quad \text{for } t \in [t_1, t_2]. \tag{1.5.5}$$

(Here $(x(t), u(t))$ is an admissible pair and $\eta(t)$ is a corresponding solution of equation (1.3.3).)

This definition implies that each Pontryagin extremal is, at the same time, an Euler extremal and, we show (see Example 1.10.2), that the set of Pontryagin extremals is a proper subset of the set of Euler extremals. We wish to mention the two special cases of control problems where the set of the Euler extremals coincides with that of the Pontryagin extremals:

1. the linear control problem which has been discussed in section 1.4,

2. nonlinear optimal control problems satisfying Assumption 1.5.1.

Note, that in the special case, namely when $F(t, x, u) = u$, the defined above Euler extremals change to integral curves of the Euler equation (see equation (1.2.2)). We show (see Chapter 4, section 4.1), that a modification of Theorem 1.5.1 remains true for the set of the Euler extremals under the following change of Assumption 1.5.1:

<u>Assumption 1.5.2</u> For each triple (t, x, η) close to $(t, x(t), \eta(t))$ there exists a unique $u(t, x, \eta)$ continuously differentiable with respect to (x, η) and measurable in t such that the following condition is satisfied:

$$H_u(t, x(t), u(t), \eta(t)) = 0 \quad \text{for } t \in [t_1, t_2].$$

The promised "true" generalized Jacobi necessary condition is presented now as follows:

<u>Theorem 1.5.2</u> If the generalized conditions of the calculus of variations hold along the chosen Euler extremal trajectory $x(t)$, then the set of the branching points of $x(t)$ coincides with that of the linear control system (1.5.1).

In particular, if $F(t, x, u) = u$, then $A(t) = 0$ and $B(t) = I$. The attainable set of the linear control system

$$\dot{z}(t) = v(t)$$

s the state space R^n itself for each $t \in (t_1, t_2]$. This implies that *the open interval (t_1, t_2) contains no branching points of $x(t)$*. Then, once again we obtain the Jacobi necessary condition.

Analogy between the conjugate and branching points.

We wish now to discuss the connection between conjugate points in the calculus of variations and branching points in the control theory. First consider the simplest problem of the calculus of variations with a strictly convex cost functional. Namely, we suppose that $f(t, x, \dot{x})$ is a strictly convex function in $(x, \dot{x})$. Consider an extremal $x(t)$. Let r be conjugate to the point t_1, i.e. there exists a sequence of real numbers $\{ r_i \}$, a sequence $\{ y_i(t) \}$ of neighboring extremals different from $x(t)$ such that

$$\lim_{i \to \infty} |y_i - x|_{D_1[t_1, t_2]} = 0, \ y_i(t_1) = x(t_1), \ y_i(r_i) = x(r_i) \text{ and } \lim_{i \to \infty} r_i = r.$$

The convexity of $f(t, x, \dot{x})$ in $(x, \dot{x})$ implies that

$$y_i(t) = x(t) \text{ on } [t_1, r_i] \text{ and } y_i(t) \neq x(t) \text{ on } (r_i, t_2].$$

Namely, the points r_i are actually the branching points of $x(t)$. The same conclusion is obtained if we substitute the condition of convexity of $f(t, x, \dot{x})$ by Hypothesis 1.2.1, which guarantees the local convexity of the Hamiltonian.

In the light of this arguments we can summarize the above discussion as follows: In the classical calculus of variations people actually tried to show the nonexistence of branching points. They did succeed in that. The only reason for this success is that the dimension of the attainable set of the linear control system $\dot{x}(t) = u(t)$ is a constant over (t_1, t_2).

Link to the Hilbert invariant integral and the Weierstrass necessary condition.

As it was promised in section 1.3, we will show now a possible extension of the constructions presented in Remark 1.2.1 in some special particular case.

The dimension of the attainable set of a general linear non autonomous control system is not necessary a constant over (t_1, t_2). Hence there no reason at all for absence of branching points. But, in spite of this fact, there are only finite number of such points which are determined by the attainable set.

Consider an extremal triple $(x(t), u(t), \eta(t))$ that satisfies Hypothesis 1.3.1 and Assumption 1.5.1. Suppose that $dim AT(t_1, t) = n$ for each $t \in (t_1, t_2)$ (here n is the dimension of the state space). In this case for some positive δ there exists a neighborhood O_δ of the curve $(t, x(t))$, such that for each $(s, y) \in O_\delta$; i.e. $|(s, x(s)) - (s, y)| < \delta$, there exists a unique solution $(x(t; \xi), \eta(t; \xi))$ of the Hamiltonian system (1.5.1), (1.5.2) with $x(t_1; \xi) = x_1$ and $x(s; \xi) = y$. Define the vector function $\varphi(s, y)$ by the following relation:

$$\varphi(s, y) = \eta(s; \xi).$$

Then, the condition $AT(t_1, t) = n$ enables, for example, the following:

the construction of a field of extremals of the control problem $(1.3.1) - (1.3.2)$; namely to derive an ordinary differential equation

$$\dot{x}(t) = \psi(t, x(t)) \tag{1.5.6}$$

such that solutions of $(1.5.6)$ are the extremals of $(1.3.1)$, $(1.3.2)$. This field of extremals makes it possible to define the Hilbert invariant integral for the optimal control problem $(1.3.1) - (1.3.2)$ as follows:

$$\int_{\Gamma} -H dt + \sum_{i=1}^{n} \eta_i dx_i.$$

The Weierstrass E–function for optimal control problem $(1.3.1) - (1.3.2)$ can also be defined by the following relation:

$$E(t, y, u, w) = f(t, y, w) - f(t, y, u) - \varphi(t, y)\{F(t, y, w) - F(t, y, u)\},$$

and then, one can obtain sufficient conditions for a strong extremum and to extend the Weierstrass necessary conditions for a strong extremum. Surprisingly this construction can be extended for much more general situations.

<u>Remark 1.5.1</u> As it has been already mentioned in section 1.4, in this work we actually consider the modification of the Pontryagin Maximum Principle. Namely, we look at the extremals which are generated by the solutions of the adjoint equation $(1.3.3)$, where η_0 is supposed to be -1. In the case of the linear control equation $(1.4.1)$ the sets of extremals which are generated by both necessary conditions coincide. However, for the nonlinear system $(1.3.1)$ it might well be the case, that the set of "true" Pontryagin extremals is bigger than that of the modification. We illustrate this statement by the following example:

<u>Example 1.5.1</u>

Consider the nonlinear control system:

$$\frac{d}{dt}x(t) = \varphi(u(t)), \quad x(0) = 0$$

with the cost functional

$$c(x, u) = \int_0^1 u^2(t) dt$$

where x, u are scalars, and $\varphi(u)$ is given as follows:

Define, first, $\varphi(u)$ on the interval $[-1, 1]$ by the relations:

$$\varphi(u) = 2u - u^2 \text{ if } 0 \le u \le 1, \quad \varphi(u) = 2u + u^2 \text{ if } -1 \le u \le 0.$$

Extend now $\varphi(u)$ on the whole real line in a smooth way such that $|\varphi(u)| \le 1$. The modification of the Pontryagin Maximum Principle generates the control function $u(t, x, \eta)$ which is given by

$$u(t, x, \eta) = \frac{\eta}{1 + \eta} \text{ for } \eta \ge 0 \quad \text{and} \quad u(t, x, \eta) = \frac{\eta}{1 - \eta} \text{ for } \eta \le 0.$$

The extremal trajectories $x(t)$ with $x(0) = 0$ have the form:

$$x(t) = \frac{\eta}{1+\eta}t \text{ for } \eta \geq 0 \quad \text{and} \quad x(t) = \frac{\eta}{1-\eta}t \text{ for } \eta \leq 0.$$

Namely, we miss two extremal trajectories $x_1(t) = t$ and $x_2(t) = -t$, which are, beyond any doubt, generated by the Pontryagin Maximum Principle. Hence, in this example the set of "true" Pontryagin extremals is bigger than that of the accepted modification.

We wish to emphasize that in this work our goal is the description of the branching of extremals which are generated by the Pontryagin Maximum Principle with $\eta_0 = -1$. However, under the Hypotheses which are accepted in the study one can easily describe the branching of "true" Pontryagin extremals by using the technique which has been developed in the section. The results which can be obtained, are similar to results obtained in this section. For this reason we do not discuss this subject in the study.

§1.6 Control systems with multidimensional performance index

The next subject which is considered in this work is the branching of extremals in control systems with multidimensional performance index. Namely, we consider the optimal control system (1.3.1) with the following k-dimensional vector cost:

$$c(x, u) = (\ c_1(x, u), \ldots, c_k(x, u)\ ), \text{ each } c_i(x, u) = \int_{t_1}^{t_2} f_i(t, x(t), u(t))dt. \tag{1.6.1}$$

In this case, namely where there are k different criteria for comparing the performance of the controls there is a conceptual difficulty in the definition of an optimal control and an optimal trajectory. We present the definitions of these objects following the terminology being used e.g., by Goffin and Haurie in [7].

Definition of extremals.

Following the notations being used e.g., by Smale in [20] we say that a point c^* which belongs to a subset C in R^k is a Pareto point of C if there is no $c \in C$ such that

$$c_i^* \geq c_i \text{ for each } i, \text{ and } c_j^* > c_j \text{ for some } j.$$

A control $u(t)$ with a corresponding trajectory $x(t)$ will be called a Pareto optimum on $[t_1, t_2]$ if there is no admissible control $w(t)$ with a corresponding trajectory $y(t)$ such that:

$$y(t_1) = x(t_1), \ y(t_2) = x(t_2) \quad \text{and}$$

$$c_i(x, u) \geq c_i(y, w) \text{ for all } i \text{ and } c_j(x, u) > c_j(y, w) \text{ for some } j.$$

In the case of a control system with a scalar cost the main object of the investigation is usually the properties of extremal trajectories, namely the trajectories which satisfy some necessary conditions. In our work we use the Pontryagin Maximum Principle as the necessary condition. We wish now to introduce a notion of extremal for the case of an optimal control system with a vector cost. In the next few lines we will justify the forthcoming notion.

Assume for a moment that we deal with a linear control system with a convex vector cost, namely where each $f_i(t, x, u)$ is convex in (x, u). Consider a Pareto pair $(x(t), u(t))$. Denote by $O(t_1, t_2)$ the set of all admissible pairs $(y(t), w(t))$ such that the following relation holds:

$$y(t_1) = x(t_1) \text{ and } y(t_2) = x(t_2).$$

Let

$$\mathbf{C} = \{\, c \mid c \in R^k, \text{ there exists } (y(t), w(t)) \in O(t_1, t_2), \text{ such that } c_i \geq c_i(y, w) \text{ each } i \,\},$$

namely $c(y, w)$ is a vector cost which one has to pay in order to transfer the system from the initial point $x(t_1)$ to the terminal point $x(t_2)$ by using a control $w(t)$, and $\mathbf{C}$ is a set of all Pareto inferior costs attached to $x(t_2)$. As we show in the study (see Chapter 5, section 5.1), in this particular case of the linear control equation the defined above cost set $\mathbf{C}$ is a closed and convex set, and $c(x, u)$ it's Pareto point. Hence there exists a positive support vector $\mu = (\mu_1, ..., \mu_k)$ to $\mathbf{C}$ at $c(x, u)$, i.e.

$$\mu_i \geq 0 \text{ for all } i, \ \mu_j > 0 \text{ for some } j$$

$$\text{such that} \quad \sum_{i=1}^{k} \mu_i c_i(x, u) \leq \sum_{i=1}^{k} \mu_i c_i(y, w) \quad \text{for each } (y, w) \in O(t_1, t_2).$$

This implies that $x(t)$ is an optimal trajectory of the linear control system with the following, so-called, μ-scalar cost:

$$c_\mu(x, u) = \sum_{i=1}^{k} \mu_i c_i(x, u), \ \mu_i \geq 0 \text{ for all } i, \ \mu_j > 0 \text{ for some } j. \tag{1.6.2}$$

This consideration helps us to define an extremal trajectory of the original nonlinear control system with k-dimensional cost, as an extremal of the same control system with some μ-scalar cost. It turns out that in this situation an extremal trajectory appears together with a positive support vector, a control function, and a solution of the corresponding adjoint equation. This encourages us to introduce the following definition:

Definition 1.6.1 A quadruple $(x(t), u(t), \eta(t), \mu)$ is an extremal quadruple of (1.3.1), (1.6.1) if a triple $(x(t), u(t), \eta(t))$ is an extremal triple the optimal control problem (1.3.1), (1.6.2).

<u>Definition 1.6.2</u> The Hamiltonian H of the optimal control problem $(1.3.1), (1.6.1)$ is defined as follows:

$$H(t, x, u, \eta, \mu) = -\sum_{i=1}^{k} \mu_i f_i(t, x, u) + \eta F(t, x, u).$$

It could happen in this case, in contrast with the case of a control system with a scalar cost, that there exist two different controls $u(t)$ and $w(t)$ which generate the same extremal trajectory $x(t)$ such that $c(x, u)$ is different from $c(x, w)$ (see Example 1.10.3). In order to distinguish between such pairs $(x(t), u(t))$ and $(x(t), w(t))$ the notions of *extended trajectory* and *extended extremal trajectory* are introduced as follows:

<u>Definition 1.6.3</u> Consider a control function $u(t)$ and a generated trajectory $x(t)$. An *extended trajectory* of the system $(1.3.1) - (1.6.1)$ is a pair $(c_u(t), x(t))$, where $c_u(t)$ is a solution of the following differential equation:

$$\frac{d}{dt}c_u(t) = (f_1(t, x(t), u(t)), ..., f_k(t, x(t), u(t))).$$

Namely, an extended trajectory is, roughly speaking, simply a pair $(cost, position)$ over $[t_1, t_2]$.

<u>Definition 1.6.4</u> An extended trajectory $(c_u(t), x(t))$ is an *extended extremal trajectory* if $x(t)$ is an extremal trajectory of $(1.3.1) - (1.6.1)$.

<u>Mathematical framework.</u>

Our purpose changes, therefore, to the investigation of branching points of extended extremal trajectories in the control problem under consideration. The underlying idea is to reduce the original control problem to the family P of the scalar control problems P_μ, where P_μ is the optimal control problem $(1.3.1), (1.6.2)$. Then the ordinary optimization machinery is applicable separately to each control problem of the constructed family P. On the other hand we intend to describe intersections of extended extremal trajectories of different scalar control problems. Namely, let $x(t)$ be an extremal trajectory of P_μ, we wish to know under what conditions there exists an extremal trajectory $y(t)$ which belongs to P_λ and $\sigma \in [t_1, t_2]$, such that $(c_w(t), y(t))$ is a neighboring extended extremal trajectory and

$$(c_u(t), x(t)) = (c_w(t), y(t)) \text{ on } [t_1, \sigma] \text{ and } (c_u(t), x(t)) \neq (c_w(t), y(t)) \text{ on } (\sigma, t_2].$$

We wish to mention, that in this case an extended extremal trajectory $(c_w(t), y(t))$ is a neighboring extended extremal trajectory if

$$\max_{t_1 \leq t \leq t_2} |(c_u(t), x(t), \eta(t), \mu) - (c_w(t), y(t), \xi(t), \lambda)| < \epsilon^{**}.$$

The value of the ϵ^{**} will be determined in Chapter 6 (see Definition 6.3.1). Such a branching of the neighboring extended extremal trajectories out of $(c_u(t), x(t))$ is closely related to the dimension of the cost set which we define as follows:

<u>Definition 1.6.5</u> The cost set at time s for the control problem (1.3.1), (1.6.1) and extended extremal trajectory $(c_u(t), x(t))$ is the set of all points $c \in R^k$, such that for some extended extremal trajectory $(c_w(t), y(t))$ satisfying the conditions

$$(c_w(t_1), y(t_1)) = (c_u(t_1), x(t_1))$$

the relation $(c_w(s), y(s)) = (c, x(s))$ holds. We shall denote this set by $CS(t_1, s)$.

Note, that in particular case of a linear control system with a convex vector cost, i.e., when each $f_i(t, x, u)$ is a convex function in (x, u), the cost set is the set of all Pareto costs attached to $x(s)$.

We introduce now the modifications of the Hypothesis 1.3.1 and Assumption 1.5.1 as follows:

<u>Hypothesis 1.6.1</u>

1. There exists a positive scalar m such that for each vector $v \in R^m$

$$v^* H_{uu}(t, x(t), u(t), \eta(t), \mu)v \leq -m|v|^2 \text{ along } x(t),$$

and $H_{uu}(t, x(t), u(t), \eta(t), \mu)$ is uniformly continuous with respect to t. Namely, for each positive ρ there exists a positive δ such that

$$|H_{uu}(t, x(t), u(t), \eta(t), \mu) - H_{uu}(t, y, w, \xi, \lambda)| < \rho$$

provided

$$|(t, x(t), u(t), \eta(t), \mu) - (t, y, w, \xi, \lambda)| < \delta.$$

2. $\int_{t_1}^{t_2} \begin{pmatrix} z(t) \\ v(t) \end{pmatrix}^* \begin{pmatrix} H_{xx}(t, x(t), u(t), \eta(t), \mu) & H_{ux}(t, x(t), u(t), \eta(t), \mu) \\ H_{xu}(t, x(t), u(t), \eta(t), \mu) & H_{uu}(t, x(t), u(t), \eta(t), \mu) \end{pmatrix} \begin{pmatrix} z(t) \\ v(t) \end{pmatrix} dt$
is negative definite for each $v(t)$, $z(t)$ such that

$$z(t) = \int_{t_1}^{t} \Phi(t, \sigma)B(\sigma)v(\sigma)d\sigma, \ z(t_2) = 0,$$

where $A(t)$, $B(t)$ are Lebesgue integrable matrices given by $A(t) = \frac{\partial F}{\partial x}(t, x(t), u(t))$, $B(t) = \frac{\partial F}{\partial u}(t, x(t), u(t))$, and $\Phi(t, t_1)$ is the transition matrix of $\frac{d}{dt}\phi = A(t)\phi$, i.e. $\phi(t) = \Phi(t, t_1)x_1$ is the solution of the equation $\frac{d}{dt}\phi = A(t)\phi$ with $\phi(t_1) = x_1$.

<u>Assumption 1.6.1</u> For each quadruple (t, x, η, λ) close to $(t, x(t), \eta(t), \mu)$ there exists a unique $u(t, x, \eta, \lambda)$ continuously differentiable with respect to (x, η, λ) and measurable in t such that

$$H(t, x, u(t, x, \eta, \lambda), \lambda) = \max_u H(t, x, u, \lambda) \text{ on } [t_1, t_2].$$

We show (see Chapter 5, section 5.2), that under these conditions the cost set of the extended extremal trajectory $(c_u(t), x(t))$ at time s is a differential manifold and its dimension is a non decreasing step function of time bounded by $k - 1$. Hence there exists a finite partition

$$t_1 = \sigma_0 < \sigma_1 < ... < \sigma_q < \sigma_{q+1} = t_2 \qquad (1.6.3)$$

of the time interval $[t_1, t_2]$ such, that each σ_j, $j = 1, ..., q$, is a point of discontinuity of this step function.

Definition 1.6.6 We denote the tangent space to the cost set $CS(t_1, s)$ at $c_u(s)$ by $W(t_1, s)$.

Remark 1.6.1 Note, that in the scalar case the dimension of the cost set of each extended extremal trajectory is zero. Namely, the cost functional does not affect the structure of the branching. Hence, once again, we obtain one of the conclusions of Corollary 1.4.1.

Existence of the branching points.

As it was shown in section 1.5, existence of a set of support vectors to the extended attainable set is the reason for the possibility of appearance of branching points. In order to investigate branching points of $(c_u(t), x(t))$ we first define the extended attainable set of control system (1.3.1) with cost (1.6.1), then introduce a tangent space $\mathcal{V}(t_1, r)$ to the extended attainable set at $(c_u(r), x(r))$. This tangent space $\mathcal{V}(t_1, r)$ characterizes support vectors to the extended attainable set at $(c_u(r), x(r))$.

Definition 1.6.7 The extended attainable set at time s for the control system (1.3.1) and the vector cost functional (1.6.1) with initial point (t_1, c_1, x_1), where $c_1 \in R^k$, $x_1 \in R^n$, is the set of all points $(c, x) \in R^k \times R^n$ such that for some extended trajectory $(c_w(t), y(t))$ satisfying the condition $(c_w(t_1), y(t_1)) = (c_1, x_1)$ the relations

$$y(s) = x, \text{ and } [c_w(s)]_i \leq c_i \text{ each } i \text{ hold.}$$

Introduce now $n \times m$ and $k \times m$ matrices $\Lambda_1(t)$ and $\Lambda_2(t)$ which will define the family of tangent spaces $\mathcal{V}(t_1, r)$, as follows:

$$\Lambda_1(t) = \Phi(t_1, t) B(t)$$

and

$$\Lambda_2(t) = \begin{pmatrix} (\int_{t_1}^t \frac{\partial f_1}{\partial x}(\sigma, x(\sigma), u(\sigma)) \Phi(\sigma, t) d\sigma) B(t) - \frac{\partial f_1}{\partial u}(t, x(t), u(t)) \\ ...... \\ ...... \\ (\int_{t_1}^t \frac{\partial f_k}{\partial x}(\sigma, x(\sigma), u(\sigma)) \Phi(\sigma, t) d\sigma) B(t) - \frac{\partial f_k}{\partial u}(t, x(t), u(t)) \end{pmatrix}$$

The space $\mathcal{V}(t_1, r)$ which is defined next, is a tangent space to the extended attainable set at $(c_u(r), x(r))$.

Definition 1.6.8 For each real number r define the vector spaces $\mathcal{V}(t_1, r)$ and $\mathcal{V}(t_1, r^+)$ as follows:

$$\mathcal{V}(t_1, r) = Sp\{ (\int_{t_1}^r \Lambda_1(t) v(t) dt, \int_{t_1}^r \Lambda_2(t) v(t) dt) \mid v \in U[t_1, r] \},$$

$$\mathcal{V}(t_1, r^+) = \bigcap_{h > 0} \mathcal{V}(t_1, r + h).$$

The structure of the branching points in this case is more complicated than that of the branching points in systems with a scalar cost. For example, the branching points of an extremal trajectory $x(t)$, even in the linear control problem, depend on the equation of the system, the cost functional and the chosen extremal $x(t)$ (see Example 1.10.4).

The branching points of the extended extremal trajectory $(c_u(t), x(t))$ are exactly the points of discontinuity of $dim \mathcal{V}(t_1, r)$. On the other hand it is shown in the study (see Chapter 5, section 5.2), that

$$\mathcal{V}(t_1, r) = V(t_1, r) \oplus W(t_1, r),$$

where $\oplus$ denotes the direct sum of subspaces in the real Euclidean space. The next statement characterizes the branching points of the extended extremal trajectory $(c_u(t), x(t))$:

Theorem 1.6.1 The set of branching points of the extended extremal trajectory $(c_u(t), x(t))$ is exactly the set $\{r_i\} \bigcup \{\sigma_j\}$. The number of branching points of the extended extremal trajectory is still finite and does not exceed $n + k - 1$, where n is the dimension of the state space and k is the dimension of the "cost" space. (Here $\{\sigma_j\}$ is the partition (1.6.3) and $\{r_i\}$ are the points of discontinuity of the dimension of the attainable set, see (1.5.3); where $A(t) = \frac{\partial F}{\partial x}(t, x(t), u(t))$ and $B(t) = \frac{\partial F}{\partial u}(t, x(t), u(t))$.)

§1.7 Optimal control problems with constraints

In the last chapter of the study we continue the description of branching points in the optimal control problem (1.3.1), (1.3.2). In contract with the preceding considerations we impose there a further restrictions or constraints. We deal with the following two main cases:

1. smooth constraints, namely constraints defined by equations of the type

$$G_i(t, x(t), u(t)) = 0, \tag{1.7.1}$$

2. inequality constraints, i.e. constraints defined by the inequalities

$$G_i(t, x(t), u(t)) \leq 0. \tag{1.7.2}$$

In both cases $G_i(t, x, u)$ is assumed to be differentiable with respect to (x, u) and measurable n t.

In the first case we manage to reduce our problem to the searching of branching points of extremals in a control problem without constraints. The structure of branching in the second ase is essentially different from that of the first one. We wish to mention that in the first case or a given pair (t, x) the set $\{ u \mid G_i(t, x, u) = 0 \}$ is a smooth manifold without boundary, in contrast with the case of inequality constraints, where the corresponding set may be a manifold with boundary, or even a convex set, like a cube. This remark helps to feel the reasons for the difference between the structures of branching.

Smooth constraints.

We derive now the condition for matching of extremals. Let $(y(t), w(t), \xi(t))$ be an extremal triple different from $(x(t), u(t), \eta(t))$ such that the relation

$$(x(t), u(t)) = (y(t), w(t)) \text{ holds on an initial subinterval } [t_1, r]. \tag{1.7.3}$$

The Pontryagin Maximum Principle implies existence of the Lagrangian multipliers $\lambda(t)$ and $\mu(t)$ such that

$$\frac{\partial}{\partial u}\{-f(t, x(t), u(t)) + \eta(t)F(t, x(t), u(t)) + \lambda(t)G(t, x(t), u(t))\} = 0$$

and

$$\frac{\partial}{\partial u}\{-f(t, y(t), w(t)) + \xi(t)F(t, y(t), w(t)) + \mu(t)G(t, y(t), w(t))\} = 0.$$

Denote $\frac{\partial F}{\partial u}(t, x(t), u(t))$ by $B(t)$ and, by using two last equations, obtain the following:

$$\Delta(t)B(t) = -(\mu(t) - \lambda(t))\frac{\partial G}{\partial u}(t, x(t), u(t)) \text{ on } [t_1, r].$$

Here $\Delta(t) = \xi(t) - \eta(t)$. Denote by $C(t)$ the orthogonal projection of R^m on

$$Sp\{ \frac{\partial G}{\partial u}(t, x(t), u(t)) \}^{\perp},$$

and obtain:

$$\Delta(t)B(t)C(t)z = -(\mu(t) - \lambda(t))\frac{\partial G}{\partial u}(t, x(t), u(t))C(t)z = 0, \tag{1.7.4}$$

for each $z \in R^m$. By using (1.7.3) and (1.3.3) one easily can show that $\Delta(t) = \Delta(t_1)\Phi(t_1, t)$, where $\Phi(t, t_1)$ is a fundamental solution of the linear differential equation $\frac{d}{dt}x = A(t)x$ with initial condition $\Phi(t_1, t_1) = I$. We denote here $\frac{\partial F}{\partial x}(t, x(t), u(t))$ by $A(t)$.

In order to describe the branching point we need to introduce the formal definition of the space of admissible controls of the following linear control system on a time interval $[\tau, s]$:

$$\dot{\phi}(t) = A(t)\phi(t) + B(t)C(t)v(t). \tag{1.7.5}$$

__Definition 1.7.1__ Let $U_c(\tau, s)$ be the set of all measurable functions defined on interval $[\tau, s]$ with range in R^m, such that for each $v \in U_c(\tau, s)$ there exists an absolutely continuous function $\phi(t)$ defined on $[\tau, s]$ with range in R^n which, is a solution of the differential equation (1.7.5) on $[\tau, s]$.

Rewrite now the relation (1.7.4) as follows:

$$\Delta(t_1)\int_{t_1}^{r} \Phi(t_1, t)B(t)C(t)v(t)dt = 0 \text{ for each } v \in U_c[t_1, r]. \tag{1.7.6}$$

We show (see Chapter 7, section 7.3), that the relation (1.7.6) is the necessary and sufficient condition for matching of the extremals. This condition yields the main result of the subsection, as follows:

Theorem 1.7.1 The set of the branching points of the extremal trajectory $x(t)$ coincides with that of the linear control system (1.7.5).

Inequality constraints.

In the remainder of the section we investigate the structure of the branching in the optimal control problem with the inequality constraints. We wish to mention that here the Pontryagin control function $u(t, x, \eta)$ is not differentiable even in the simple case where the control system is $\dot{x}(t) = u(t)$, the cost functional is given by $c(x, u) = \int_{t_1}^{t_2} u^2(t)dt$ and the constraints are, e.g.,

$$G_1(t, x, u) = u + 1, \quad G_2(t, x, u) = u - 1.$$

But, fortunately, in many cases, this function is continuous and Lipschitz. These facts still allow us to consider the extremals as the solutions of the Hamiltonian system (1.5.1), (1.5.2). In order to derive the necessary and sufficient conditions for branching of an extremal trajectory we develop now the necessary series of auxiliary notions and relations. Let $(x(t), u(t), \eta(t))$ be an extremal triple of the optimal control problem (1.3.1), (1.3.2) with the set of convex constraints

$$G(t, x, u) = (\ G_1(t, x, u), ..., G_k(t, x, u)\ ).$$

Then

$$H_u(t, x(t), u(t), \eta(t)) \in co(t) \text{ on } [t_1, t_2],$$

where by $co(t)$ we denote the following closed, convex cone

$$\{\ z\ |\ z \in R^m, z = \sum_{i=1}^{k} \alpha_i \frac{\partial G_i}{\partial u}(t, x(t), u(t)),\ \alpha_i \geq 0 \text{ for all } i,\ \alpha_j = 0 \text{ if } G_j(t, x(t), u(t)) < 0\ \}.$$

Suppose that $(y(t), w(t), \xi(t))$ is an extremal triple such that the relation

$$(x(t), u(t)) = (y(t), w(t))$$

holds on $[t_1, r]$. This implies that

$$H_u(t, y(t), w(t), \mu(t)) \in co(t) \text{ on } [t_1, r].$$

In other words

$$H_u(t, y(t), w(t), \mu(t)) + (\xi(t_1) - \eta(t_1))\Phi(t_1, t)B(t) \in co(t) \text{ on } [t_1, r].$$

We introduce now the convex set $C(t_1, s)$ which determines branching points of the extremal trajectory $x(t)$.

Definition 1.7.2 Let

$$C(t_1, s) = \{\ c\ |\ c \in R^n,\ H_u(t, x(t), u(t), \eta(t)) + c\Phi(t_1, t)B(t) \in co(t),\ t \in [t_1, s]\ \}.$$

We wish to emphasize two evident, but extremely useful, properties of $C(t_1, s)$:

1. $C(t_1, s)$ is a convex, closed set for each $s \in [t_1, t_2]$,
2. $C(t_1, s_1) \supset C(t_1, s_2)$ provided $s_1 \leq s_2$.

The set $C(t_1, s)$, as well as the set $V(t_1, s)^\perp$ in the case of the control problems without constraints, is a set of "supports" to the attainable set at time s. In the light of this remark it is easy to understand the appearance of the symbol $\bigcup$ in the next definition, instead of $\bigcap$ in the corresponding definition of $V(t_1, s)$.

Definition 1.7.3 We define the convex set $C(t_1, r^+)$ as follows:

$$C(t_1, r^+) = \bigcup_{h>0} C(t_1, r + h).$$

The one parameter family of convex, closed sets $C(t_1, s)$ characterizes the branching points of $x(t)$, namely the branching points of $x(t)$ are exactly the points where $C(t_1, s)$ decrease, in other words:

Theorem 1.7.2 A point r is a branching point of the extremal trajectory $x(t)$ if and only if

$$C(t_1, r^+) \text{ is a proper subset of } C(t_1, r).$$

In particular, the "continuous" branching is possible in this situation, i.e. there exists an extremal trajectory $x(t)$ such that for each s which belongs to $[t_1, t_2]$ there exists an extremal $x_s(t)$ which branches out of $x(t)$ at s, that is

$$x(t) = x_s(t) \text{ if } t_1 \leq t \leq s \text{ and } x(t) \neq x_s(t) \text{ if } s < t \leq t_2.$$

(See Example 1.10.5.)

§1.8 Branching pairs

The present section is devoted to a more general phenomenon of branching. Namely, consider extremal trajectories $x(t)$, $y(t)$ which coincide on some subinterval $[l, r]$ of the time interval $[t_1, t_2$

and differ outside of $[l, r]$. It must be mentioned that here $[l, r]$ is not necessary an initial subinterval of $[t_1, t_2]$, i.e. l may be different from t_1. We call such a pair $\{l, r\}$ a branching pair of the extremal trajectory $x(t)$, namely:

Definition 1.8.1 A pair of real numbers $\{l, r\}$ such that $t_1 \leq l < r \leq t_2$ is a branching pair of the extremal trajectory $x(t)$ if there exists an extremal trajectory $y(t)$ in a family of neighboring trajectories such that

$$x(t) = y(t) \text{ on } [l, r], \quad x(t) \neq y(t) \text{ on } [t_1, l), \text{ and } x(t) \neq y(t) \text{ on } (r, t_2].$$

In this section we investigate branching pairs of extremal trajectories in four optimal control problems, that have been discussed in the preceding sections. The analysis in this section is, therefore, contained in four parts, each of them is devoted to description of the phenomenon in one of the types of control problems. A central role in determination of the branching pairs of extremal trajectories in optimal control problems under consideration belongs to the sets $V(l, r)$, $\mathcal{V}(l, r)$ and $C(l, r)$. (See Definition 1.4.2, Definition 1.6.8 and Definition 1.7.2.) Note that now, when l does not necessary coincide with t_1, each one of the three sets forms actually a two parameter family of sets.

Three families of sets which are defined below will be extremely important in description of branching pairs.

Definition 1.8.2 We define the three following families of sets $V(r, l^-)$, $\mathcal{V}(r, l^-)$ and $C(r, l^-)$ as follows:

$$V(r, l^-) = \bigcap_{h>0} V(r, l - h),$$

$$\mathcal{V}(r, l^-) = \bigcap_{h>0} \mathcal{V}(r, l - h),$$

$$C(r, l^-) = \bigcup_{h>0} C(r, l - h).$$

In the following subsections we investigate branching pairs of extremals in the different optimal control problems. The basic difference between the sets of branching points and branching pairs is pointed out, and the distinction between the sets of branching pairs in control problems of various types is clarified.

Linear systems with convex cost.

The linear control system (1.4.1) with the scalar cost (1.3.2), where $f(t, x, u)$ is supposed to be a convex function of (x, u), is considered in this subsection. A pair $\{l, r\}$ is a branching pair if and only if the following condition satisfied:

$V(l, r)$ is a proper subspace of $V(l, r^+)$ and $V(r, l)$ is a proper subspace of $V(r, l^-)$.

The following theorem is the main result of the subsection:

Theorem 1.8.1 The set of the branching pairs of an extremal trajectory $x(t)$ depends on the linear control equation (1.4.1) (namely the matrices $A(t)$ and $B(t)$) only. It does not depend on the cost functional or the chosen extremal $x(t)$. This set is at most countable. (The detailed proof of this result is presented in Chapter 3, section 3.3.)

The structure of the branching pairs is appreciably different from that of the branching points. We complete this subsection with the following example of a linear control problem with a countable set of branching pairs.

Example 1.8.1 A linear control system with a countable set of branching pairs.

Let $g(t; a, b)$ be a real valued function of a real variable t defined on an interval $[a, b]$ as follows:

$$t - a \quad \text{for } a \leq t \leq \frac{a+b}{2}, \quad \frac{b-a}{2} - t \quad \text{for } \frac{a+b}{2} < t \leq b.$$

Note that $g(t; a, b)$ is continuous in t on (a, b), $\dot{g} = 1$ on $(a, \frac{a+b}{2})$, $\dot{g} = -1$ on $(\frac{a+b}{2}, b)$, and $g(a; a, b) = g(b; a, b) = 0$.

Next, consider the linear control system

$$\dot{x} = B(t)u(t)$$

with the quadratic cost

$$c(x, u) = \int_0^1 u^2(t)dt$$

where x and u are scalars, and $B(t)$ is a scalar function defined below.

For each $k = 0, 1, ...$ define the set of intervals as follows:

$$I_{k1} = (2/2^{3k+1}, 1/2^{3k}], \quad I_{k2} = [2/2^{3k+2}, 1/2^{3k+1}], \quad I_{k3} = [2/2^{3k+3}, 1/2^{3k+2}).$$

Note that

$$\bigcup_{k=0}^{\infty} (I_{k1} \cup I_{k2} \cup I_{k3}) = (0, 1].$$

Define $B(t)$ as follows:

$B(t) = 0$ for $t = 0$ or $t \in I_{k2}$,
$B(t) = g(t; 1/2^{3k+1}, 1/2^{3k})$ for $t \in I_{k1}$,
$B(t) = g(t; 1/2^{3k+3}, 1/2^{3k+2})$ for $t \in I_{k3}$.

For each extremal triple $(x(t), u(t), \eta(t))$ of this system, the following conditions hold:

$$\dot{\eta}(t) = 0 \text{ and } \max_u \{-u^2 + \eta(t)B(t)u\} = -u^2(t) + \eta(t)B(t)u(t).$$

This implies that:

1. $\eta(t)$ is a constant which we denote by η,
2. $u(t) = \frac{1}{2}\eta B(t)$.

Choose an extremal triple $(x(t), u(t), \eta(t))$. For each integer $k = 0, 1, \dots$ we wish to define an extremal triple $(x_k(t), u_k(t), \eta_k(t))$ such that:

1. $x_k(t) = x(t)$ if $t \in I_{k2}$,
2. $x_k(t) \neq x(t)$ otherwise.

Define: $\eta_k(t)$ to be a constant η_k different from η,

$u_k(t) = \frac{1}{2}\eta_k B(t)$,

$x_k(t) = x(t)$ for $t \in I_{k2}$ and $x_k(t) = x_k(2/2^{3k+1}) + \int_{1/2^{3k+1}}^{t} \eta_k B(s)ds$ otherwise.

The defined above triples $(x_k(t), u_k(t), \eta_k(t))$ are extremal triples such, that $x_k(t)$ branches out of $x(t)$ at $\{ 1/2^{3k+1}, 1/2^{3k+2} \}$. This means that the set of branching pairs of $x(t)$ is not finite.

Branching pairs, as well as, branching points of a linear control system are determined by points of discontinuity of the attainable set. Let us consider a control system with analytic coefficients $A(t)$, $B(t)$; namely each entry of these matrices is a real analytic function of time. One can show that the dimension of the attainable set in this case is a constant. Hence, linear control systems with analytic coefficients, have no branching at all.

Note, that in the presented example $A(t)$ is an analytic, and $B(t)$ is a continuous functions of time. We wish to mention that the non smoothness of $B(t)$ is not significant for branching. By using a slight modification of the example one can easily construct a linear control system with even C^∞ coefficients, which has an infinite countable set of branching pairs.

Nonlinear systems.

Let $x(t)$ be an extremal trajectory of the optimal control problem (1.3.1), (1.3.2). Suppose that $x(t)$ satisfy Hypothesis 1.3.1 and Assumption 1.5.1. Then the branching pairs of the extremal trajectory $x(t)$ are given by the following theorem:

Theorem 1.8.2 The branching pairs of an extremal trajectory $x(t)$ are coincide with that of the following linear control system:

$$\dot{z}(t) = A(t)z(t) + B(t)v(t),$$

here $A(t) = \frac{\partial F}{\partial x}(t, x(t), u(t))$ and $B(t) = \frac{\partial F}{\partial u}(t, x(t), u(t))$.

The proof can be found in Chapter 4, section 4.5.)

Control systems with vector cost.

Consider an extended extremal trajectory $(c_u(t), x(t))$ which satisfy Hypothesis 1.6.1 and Assumption 1.6.1. The branching pairs of $(c_u(t), x(t))$ are characterized by the next two statements, as follows:

<u>Lemma 1.8.1</u> A pair $\{l, r\}$ is a branching pair of an extremal trajectory $x(t)$ if and only if the following condition holds:

$$\mathcal{V}(l, r) \text{ is a proper subspace of } \mathcal{V}(l, r^+) \text{ and } \mathcal{V}(r, l) \text{ is a proper subspace of } \mathcal{V}(r, l^-).$$

(The proof of Lemma 1.8.1 is displayed in Chapter 6, section 6.4.)

<u>Theorem 1.8.3</u> The set of the branching pairs of an extended extremal trajectory $(c_u(t), x(t))$ depends on the linear control equation (1.4.1), the cost functional (1.6.1), and the chosen extended extremal trajectory $(c_u(t), x(t))$. This set is at most infinite countable.

The proofs of these results are included in section 5.2 and section 5.3 of Chapter 5.

Optimal control problems with constraints.

We consider, first, an extremal trajectory $x(t)$ of the optimal control problem (1.3.1), (1.3.2) with the smooth constraints (1.7.1). Our main result in this case, is the following:

<u>Theorem 1.8.4</u> The set of the branching pairs of the extremal trajectory $x(t)$ coincides with that of the linear control system (1.7.5).

The last result of the section concerns the branching pairs of an extremal trajectory $x(t)$ in a control problem with inequality constraints. The key role in description of branching pairs belongs to the two parameter family of closed, convex sets $C(l, r)$.

<u>Theorem 1.8.5</u> A pair $\{l, r\}$ is a branching pair of an extremal trajectory $x(t)$ if and only if there exists a vector $c \in R^n$ such that the following conditions hold:

1. $c \in C(l, r)$,
2. $c\Phi(r, l) \in C(r, l)$,
3. $c \notin C(l, r^+)$,
4. $c\Phi(r, l) \notin C(r, l^-)$.

<u>Example 1.8.2</u>

Note that in the case of a control problem without constraints the conditions

$$V(l, r)^\perp \neq V(l, r^+)^\perp \quad \text{and} \quad V(r, l)^\perp \neq V(r, l^-)^\perp$$

are necessary and sufficient for existence of branching. Namely, as it is shown in the study (see Lemma 3.3.1 and Lemma 3.3.2 of Chapter 3), these conditions imply existence of such $c \in R^r$ that

$$c \in V(l, r)^\perp - V(l, r^+)^\perp \quad \text{and} \quad c\Phi(r, l) \in V(r, l)^\perp - V(r, l^-)^\perp.$$

We show in this example that in the present case, in contrast with the case of control problems without constraints, the conditions

$$C(l,r) \neq C(l,r^+) \quad \text{and} \quad C(r,l) \neq C(r,l^-)$$

generally do not imply existence of such c that

$$c \in C(l,r) - C(l,r^+) \quad \text{and} \quad c\Phi(r,l) \in C(r,l) - C(r,l^-).$$

Consider the linear control system:

$$\dot{x}(t) = u, \; x(0) = 0,$$

with the cost functional

$$c(x,u) = \int_0^3 u^2(t)dt$$

and two linear constraints

$$g_1(t)u(t) \leq 0, \; g_2(t)u(t) \leq 0.$$

Where $x, u \in R^2$ and the vectors $g_1(t)$, $g_2(t)$ are defined as follows:

For $0 \leq t \leq 1$ $g_1(t) = (-1,0)$, $g_2(t) = (-\sin \frac{\pi}{2}(t-1), -\cos \frac{\pi}{2}(t-1))$.

For $1 < t \leq 2$ $g_1(t) = (-1,0)$, $g_2(t) = (0,-1)$.

For $2 < t \leq 3$ $g_1(t) = (-\cos \frac{\pi}{2}(t-2), \sin \frac{\pi}{2}(t-2))$, $g_2(t) = (-1,0)$.

Consider the extremal triple $(0,0,0)$. In this case

$$C(1,2) = C(2,1) = \{ c \mid c \in R^2, \; c_1 \geq 0, \; c_2 \geq 0 \}.$$

On the other hand

$$C(1,2^+) = \{ c \mid c \in R^2, \; c_1 \geq 0, \; c_2 > 0 \}$$

and

$$C(2,1^-) = \{ c \mid c \in R^2, \; c_1 > 0, \; c_2 \geq 0 \}.$$

Namely, there is no such $c \in R^2$ that

$$c \in C(1,2), \; c \notin C(1,2^+), \; c \notin C(2,1^-).$$

§1.9 Summary

The classical theory of the calculus of variations presents the sufficiency criterion which enables to derive the Jacobi necessary condition. This condition implies the absence of branching points in problems of the calculus of variations. Branching points of extremal trajectories do occur in optimal control even under the natural generalization of the sufficiency criterion of the calculus of variations. In this work we have associated branching points with jump points of the dimensionality of the attainable set. The conclusion is that only a finite number of branching points exists, and if the dimension of the attainable set does not change, then the branching points do not exist.

Another way to describe branching points is as follows: Actually, the nested families of the sets $V(t_1,t)^{\perp}$, $\mathcal{V}(t_1,t)^{\perp}$ and $C(t_1,t)$ are responsible for the existence of the branching points. From the geometrical point of view these sets, as it was shown in section 1.4 of this Chapter, can be associated in a natural way with the sets of support vectors to the attainable sets of the corresponding control problems. It must be mentioned that we always consider the attainable sets of the linear control systems. The attainable sets are linear spaces or convex cones (in the case of control problems with constraints). In convex analysis, (see, for example Rockafellar, [16], p. 121.) the set of support vectors to a convex cone is called a polar.

This analogy encourage me to propose to call each one of the sets $V(t_1,t)^{\perp}$, $\mathcal{V}(t_1,t)^{\perp}$ and $C(t_1,t)$ the polar at the point t of the corresponding control system. We wish to emphasize, once more, that the polar, as well as the attainable set, depends only on the control equation, and does not depend on the cost functional. Then finally we can say, that a point r is a branching point if and only if for each positive h the polar at the point r is strictly included in the polar at the point $r + h$.

§1.10 Examples

Example 1.10.1

The notion of neighboring extremal (see section 1.3) plays an important role in the investigation of branching points. In this example we present a bilinear control problem with extremals $x(t)$ and $y(t)$ satisfying the relations:

$$x(0) = y(0), \quad x(\pi) = y(\pi), \quad x(t) \neq y(t) \text{ for } t \in (0, \pi).$$

Consider the following bilinear control system:

$$\dot{x}_1(t) = u(t)x_2(t), \qquad (1.10.1)$$

$$\dot{x}_2(t) = -u(t)x_1(t), \qquad (1.10.2)$$

where $u(t)$ is a scalar function.

Consider the cost functional

$$c(x, u) = \frac{1}{2} \int_0^{2\pi} u^2(t)dt. \tag{1.10.3}$$

The adjoint equation has the following form:

$$\dot{\eta}_1(t) = u(t)\eta_2(t),$$

$$\dot{\eta}_2(t) = -u(t)\eta_1(t).$$

The Hamiltonian H of this problem is given (see Definition 1.3.4) as follows:

$$H(t, x, u, \eta) = -\frac{1}{2}u^2 + (\eta_1 x_2 - \eta_2 x_1)u.$$

Note that $\eta_1(t)x_2(t) - \eta_2(t)x_1(t)$ is a constant and, therefore, $u(t) = u(0)$ for $t \in [0, 2\pi]$.

The extremal trajectories $x_\omega(t)$ of the above optimal control problem are rotations around the origin with constant angular velocities ω. For each real number ω, the extremals $x_\omega(t)$ and $x_{-\omega}(t)$ initiating at the same point $x(0)$ intersect each other at $t = \frac{\pi}{\omega}$. Assumption 1.5.1 is evidently fulfilled and we will choose ω in such a way that the extremal trajectory $x_\omega(t)$ will satisfy Hypothesis 1.3.1 on $[0, 2\pi]$.

In order to find the suitable ω, choose the following set of initial conditions:

$$x_1(0) = 1, \; x_2(0) = 0, \; \eta_1(0) = 0, \; \eta_2(0) = -1.$$

Straightforward verification would show that in this case $u(t) = 1$ and

$$x_1(t) = \cos t, \; x_2(t) = -\sin t, \; \eta_1(t) = -\sin t, \; \eta_2(t) = -\cos t.$$

The linearized system about this solution is the following:

$$\dot{z}_1(t) = z_2(t) - v(t)\sin t,$$

$$\dot{z}_2(t) = -z_1(t) - v(t)\cos t.$$

Those solutions $z(t) = (z_1(t), z_2(t))$ of this linear equation that satisfy also the boundary conditions

$$z_1(0) = z_1(2\pi) = z_2(0) = z_2(2\pi) = 0$$

 have the following form:

$$z_1(t) = -\sin t \int_0^t v(s)ds, \; z_2(t) = -\cos t \int_0^t v(s)ds. \tag{1.10.4}$$

We now wish to show that the extremal triple $(x(t), u(t), \eta(t))$ satisfies Hypothesis 1.3.1. A straightforward computation shows that:

1. $H_{uu}(t, x, u, \eta)$ is identically equal to -1,

2. $\displaystyle\int_0^{2\pi} \begin{pmatrix} z(t) \\ v(t) \end{pmatrix}^* \begin{pmatrix} H_{xx}(t, x(t), u(t), \eta(t)) & H_{ux}(t, x(t), u(t), \eta(t)) \\ H_{xu}(t, x(t), u(t), \eta(t)) & H_{uu}(t, x(t), u(t), \eta(t)) \end{pmatrix} \begin{pmatrix} z(t) \\ v(t) \end{pmatrix} dt =$

$$= -\int_0^{2\pi} v^2(t) dt$$

is negative definite for each $z(t)$ satisfying condition (1.10.4).

On the other hand, the trajectory $y(t) = (\cos t, \sin t)$ is an extremal trajectory of the optimal control problem (1.10.1), (1.10.2), (1.10.3) and

$$y(\pi) = x(\pi), \ y(0) = x(0), \text{ but } y(t) \neq x(t) \text{ for } 0 < t < \pi.$$

Note that $\mu_1(0) = 0$, $\mu_2(0) = 1$ and $\eta_1(0) = 0$, $\eta_2(0) = -1$, hence $|(x, \eta) - (y, \mu)|_{C[t_1, t_2]} \geq 2$. On the other hand, one can show that, in this case, ϵ^* which defines the extremal trajectories must be less than 1.

Example 1.10.2

The relationship between the sets of Pontryagin and Euler control functions was discussed in section 1.5. In this example we introduce a control system with different Pontryagin and Euler control functions.

Consider the nonlinear control system

$$\dot{x}(t) = \frac{1}{2} u^2(t), \quad x(0) = 0,$$

with the cost functional

$$c(x, u) = \int_0^1 \{\frac{1}{4} u^4(t) - \frac{2}{3} u^3(t) + \frac{1}{2} u^2(t)\} dt,$$

where x and u are scalars. We restrict ourselves to the region $\{(t, x, \eta); \ \eta > 1\}$. In this case we obtain **three** different continuously differentiable control functions

$$u_1(t, x, \eta) = 0, \quad u_2(t, x, \eta) = 1 + \sqrt{\eta}, \quad u_3(t, x, \eta) = 1 - \sqrt{\eta},$$

such that

$$H_u(t, x, u_i(t, x, \eta), \eta) = 0, \quad \text{for } i = 1, 2, 3.$$

On the other hand

$$\max_u H(t, x, u, \eta) = H(t, x, u_2(t, x, \eta), \eta).$$

In this example, therefore, the set of Pontryagin controls is a proper subset of the set of Euler controls.

Example 1.10.3

As it was already mentioned in section 1.6, it might well be the case, that control functions $u(t)$ and $w(t)$ of a control system with multidimensional cost, generate the same extremal trajectory $x(t)$. We present now a concrete example of such a control problem.

Consider the linear control system

$$\dot{x}(t) = u_1(t) + u_2(t), \quad x(0) = 0,$$

with a vector cost functional

$$c_1(x, u) = \int_0^2 u_1^2(t) + u_2^2(t)dt, \quad c_2(x, u) = \int_0^2 \varphi(t)u_1^2(t) + \psi(t)u_2^2(t)dt.$$

Where x is a scalar, $u \in R^2$, $\varphi(t)$, $\psi(t)$ are the real valued functions which are defined as follows:

$$\varphi(t) = 1 \text{ on } [0, 1], \quad \varphi(t) = 3/2 \text{ on } (1, 2],$$

$$\psi(t) = 1 \text{ on } [0, 1], \quad \psi(t) = 3/4 \text{ on } (1, 2].$$

The only reason for the choice of the step functions $\varphi(t)$ and $\psi(t)$ is the simplicity of the following calculations. One can construct a modification of this example even with C^∞ functions $\varphi(t)$ and $\psi(t)$. Consider an admissible control $u(t)$ and a generated trajectory $x(t)$ which are defined as follows:

$$u(t) = (1, 1) \text{ and } x(t) = 2t.$$

Note, that this pair is an extremal pair of $(1, 0)$-control problem. In this case

$$c_1(x, u) = 4, \quad c_2(x, u) = 4.25.$$

Consider now an extremal pair $(y(t), w(t))$ of $(0, 1)$-control problem which is defined as follows:

$$y(t) = 2t \text{ on } [0, 2],$$

$$w_1(t) = 1, \quad w_2(t) = 1 \text{ on } [0, 1],$$

$$w_1(t) = \frac{2}{3}, \quad w_2(t) = \frac{4}{3} \text{ on } (1, 2].$$

In this case

$$c_1(y, w) = \frac{38}{9}, \quad c_2(y, w) = 4.$$

Example 1.10.4 Pareto extremal trajectories with different sets of branching points.

Branching points of an extended extremal trajectory $(c_u(t), x(t))$ depend on a control equation, a cost functional, and the trajectory itself. This subject has been discussed in section 1.6 and now the illustration of this fact is presented as follows:

Consider the linear control system on $[-1,1]$

$$\dot{x}(t) = u(t), \text{ with } x(-1) = -1,$$

with the following vector cost functional

$$c_1(x,u) = \frac{1}{2}\int_{-1}^{1} u^2(t)dt, \quad c_2(x,u) = \frac{1}{2}\int_{-1}^{1} \varphi(t,x(t))u^2(t)dt.$$

Where x, u are scalars and $\varphi(t,x)$ is a real valued function defined as follows:

$$\varphi(t,x) = 1 + (t-x)^2 \text{ if } x > t > 0 \text{ and } \varphi(t,x) = 1 \text{ otherwise.}$$

This $\varphi(t,x)$ is chosen for the sake of computational convenience. One can construct a modification of the example even with C^∞ function $\varphi(t,x)$. Choose $\lambda_1 = \lambda_2 = 1$. The Hamiltonian H has the following form:

$$H(t,x,u,\eta,\lambda) = -\frac{1}{2}\{ u^2 + \varphi(t,x)u^2 \} + \eta u$$

$$\text{and } u(t,x,\eta,\lambda) = \frac{\eta}{1+\varphi(t,x)}.$$

Note that $dimV(-1,t) = 1$ for each $t \in (-1,1]$.

We intend, now to define the two promised extended extremal trajectories. Consider first the following initial data:

$$x_1(-1) = -1, \quad \eta_1(-1) = 1.$$

A straightforward computation would show that $x_1(t) = -1 + \frac{1}{2}(t+1)$ is a Pareto extremal trajectory. This yields that $\varphi(t,x_1(t)) = 1$ on $[-1,1]$. Hence $dimV(-1,t) = 1$ for each $t \in (-1,1]$. In accordance with Theorem 1.6.1 the set of branching points of the extended Pareto trajectory $(c_{u_1}(t), x_1(t))$ is the empty set.

Consider now the following initial conditions:

$$x_2(-1) = -1, \quad \eta_2(-1) = 4.$$

In this case $x_2(t) = -1+2(t+1)$ for $t \in [-1,0]$, in particular $x_2(0) = 1$. There exists $\epsilon > 0$ such that the relation $x_2(t) > t$ holds for each $t \in [0,\epsilon]$ and, therefore, $\varphi(t,x_2(t)) \neq 1$. This immediately implies that $dimV(-1,0^+) = 2$. Then, due to Theorem 1.6.1, there exists an extended Pareto trajectory $(c_w(t), y(t))$ which branches out of $(c_{u_2}(t), x_2(t))$ at the point 1.

Example 1.10.5

The existence of "continuous" branching was promised in section 7.1. In this example we present a linear control system with a continuum of branching points.

Consider the linear control system:

$$\dot{x}(t) = u, \quad x(0) = 0,$$

with the following set of two constraints:

$$-1 \le u, \quad u \le 1, \text{ namely } G_1(t, x, u) = u - 1, \quad G_2(t, x, u) = -u - 1.$$

Consider the cost functional

$$c(x, u) = \frac{1}{2} \int_0^1 (t + 1) u^2(t) dt$$

where x, u are scalars. In this case a control function $u(t, x, \eta)$ is the following:

$$\frac{\eta(0)}{1 + t} \quad \text{if } |\eta(0)| \le 1 + t, \quad 1 \quad \text{if } \eta(0) > 1 + t, \quad -1 \quad \text{if } \eta(0) < -(1 + t).$$

Let $\eta(0) = 2$, then $u(t, x(t), \eta(t)) = 1$ on $[0, 1]$ and $x(t) = t$ for $t \in [0, 1]$. On the other hand for each $0 \le s \le 1$ choose $\eta_s(0) = \frac{1+s}{1+t}$, then

$$x_s(t) = t \text{ for } 0 \le t \le s \text{ and } x_s(t) = s + (1 + s) \ln \left(\frac{1 + t}{1 + s} \right).$$

This implies that each $s \in [0, 1)$ is a branching point of the extremal trajectory $x(t) = t$.

Note, that the definition of $C(t_1, s)$ implies that

$$C(t_1, s) = \{ c \mid c \text{ is a scalar such that } - (1 + s) + 2 + c \ge 0 \}.$$

Hence, in accordance with Theorem 1.7.2, $C(t_1, s_1) \ne C(t_1, s_2)$ provided $s_1 \ne s_2$ and $C(t_1, s) \ne C(t_1, s^+)$ for each $s \in [0, 1)$.

Chapter 2

Branching points in linear control problems

§2.1 Foreword

An extremal trajectory of an optimal control problem is usually a good candidate for the optimal one. This is the main motivation behind beginning a search for the optimal trajectories from investigation extremal trajectories. Unfortunately a set of optimal trajectories, in general, is a proper subset of a set of extremal trajectories. However the set of optimal trajectories in the problem presented in this chapter coincides with that of extremal trajectories.

In the present chapter we deal with extremal trajectories of linear optimal control problems. In this chapter an extremal trajectory is a trajectory which satisfies the Pontryagin Maximum Principle. The phenomenon we wish to describe is as follows: Consider an extremal trajectory $x(t)$ defined over time interval $[t_1, t_2]$. Suppose that there exists an extremal $y(t)$ such that $x(t)$ and $y(t)$ coincide over an initial subinterval $[t_1, r]$ of the time interval and differ on the rest of it. Namely, $x(t) = y(t)$ for each $t \in [t_1, r]$, and $x(t) \neq y(t)$ for each $t \in (r, t_2]$. In this case we say that r is a *branching point* of the extremal trajectory $x(t)$ and $y(t)$ *branches out of* $x(t)$ at r.

In this chapter we investigate branching points of linear control systems. Our goal is to describe the structure of branching of the extremal trajectories in the control systems under consideration. Our main result is as follows: The set of the branching points of an extremal $x(t)$ is a finite set which depends on the equation of a control system only. In particular, this set does not depend on the reference optimal trajectory $x(t)$ or on a cost functional.

In the classical theory of the calculus of variations, the problem of intersection of neighboring extremals is considered. There this topic is treated under the conditions which ensure the absence of splittings or intersections of the neighboring extremals. The results obtained extend a known criterion in the calculus of variations involving the Jacobi condition for the problems with convex Lagrangians.

§2.2 Look at the solutions of the Hamiltonian system

In this section we prepare necessary tools for deriving necessary and sufficient conditions for branching. The main idea is to consider a pair $(x(t), \eta(t))$ of an extremal trajectory and a corresponding solution of the adjoint equation (1.3.3) as a solution of the given below system of ordinary differential equations (2.2.1), (2.2.2).

We would like to remind the reader, that an extremal trajectory $x(t)$ is a first component of an extremal triple $(x(t), u(t), \eta(t))$. If for each (t, x, η) there exists a unique $u(t, x, \eta)$ such that the relation (1.3.4) holds, then the substitution of $u(t, x, \eta)$ into the equations (1.4.1), (1.3.3) yields the following system of ordinary differential equations:

$$\frac{d}{dt}x(t) = A(t)x(t) + B(t)u(t, x(t), \eta(t)) \tag{2.2.1}$$

$$\frac{d}{dt}\eta(t) = \frac{\partial}{\partial x}f(t, x(t), u(t, x(t), \eta(t))) - \eta(t)A(t). \tag{2.2.2}$$

We say that the optimal control problem (1.4.1), (1.3.3) is *optimal equivalent* to the system (2.2.1), (2.2.2) if the following conditions hold:

1. each extremal triple $(x(t), u(t), \eta(t))$ of (1.4.1), (1.3.3) satisfies the three conditions displayed below:

    a. $u(t) = u(t, x(t), \eta(t))$,

    b. $x(t)$ is a solution of the equation (2.2.1),

    c. $\eta(t)$ is a solution of the equation (2.2.2)

    (we say that $(x(t), \eta(t))$ is a solution of (2.2.1), (2.2.2) generated by the extremal triple $(x(t), u(t), \eta(t))$),

2. for each solution $(x(t), \eta(t))$ of the system (2.2.1), (2.2.2) the triple $(x(t), u(t, x(t), \eta(t)), \eta(t))$ is an extremal triple of the optimal control problem (1.4.1), (1.3.3).

In the case of the linear control equation (1.4.1) with the convex cost (1.3.2), the set of the extremal trajectories coincides with that of the optimal trajectories. (We wish to emphasize that in this chapter $(x(t), u(t))$ is an optimal pair of (1.4.1), (1.3.2) if for any admissible pair $(y(t), w(t))$ with $y(t_1) = x(t_1)$ and $y(t_2) = x(t_2)$ the following condition holds:

$$c(x, u) - c(y, w) \le 0.)$$

This subject was considered partly in [4]. In the case of the linear system (1.4.1) with the convex cost of the form:

$$c(x, u) = \int_{t_1}^{t_2} h(t, u(t)) + g(t, x(t))dt$$

the existence of such a correspondence is proved in [11], p. 212. We did not find a relevant reference for our case. In the Appendix we prove the optimal equivalence of the optimal control problem (1.4.1), (1.3.3) and the system of ordinary differential equations (2.2.1), (2.2.2), and show that each extremal trajectory of (1.4.1), (1.3.3) is, at the same time an optimal trajectory. Let us mention here an important conclusion which is yielded by the results of the Appendix: If $(x(t), u(t))$ and $(y(t), w(t))$ are extremal pairs of (1.4.1), (1.3.3) such that $x(s) = y(s)$ and $x(\tau) = y(\tau)$ then $(x(t), u(t)) = (y(t), w(t))$ over $[s, \tau]$.

The last remark of this section is the following: By using the Hamiltonian one can rewrite the system (2.2.1), (2.2.2) as the Hamiltonian system (1.5.1), (1.5.2):

$$\dot{x}(t) = H_\eta$$
$$\dot{\eta}(t) = -H_x.$$

Then the main message of the section is as follows: There exists a one-to-one correspondence between extremal triples of (1.4.1), (1.3.3) and solutions of the Hamiltonian system (1.5.1), (1.5.2). In view of the fact that the set of optimal trajectories coincides with that of extremal trajectories, each solution of the system (2.2.1), (2.2.2) generates an optimal trajectory of (1.4.1), (1.3.2).

§2.3 Branching of extremals

Throughout this chapter we consider an extremal triple $(x(t), u(t), \eta(t))$. In this section we study the extremals $\{y(t)\}$ which initiate at the same point $x(t_1)$ as $x(t)$. Our goal is the description of the branching points of the extremal trajectory $x(t)$. The main result of this section is Theorem 1.4.1 which has been presented in the first chapter.

It was shown in Chapter 1 (see Lemma 1.4.1) that the condition

$$\mu(t_1) - \eta(t_1) \in V(t_1, r)$$

is necessary for the matching of extremal trajectories $x(t)$ and $y(t)$ on the initial subinterval $[t_1, r]$. In the next lemma we show that this condition is also sufficient for matching of the extremals $x(t)$ and $y(t)$.

Lemma 2.3.1 Consider an extremal triple $(y(t), w(t), \mu(t))$ such that $x(t_1) = y(t_1)$. If the relation $\eta(t_1) - \mu(t_1) \in V(t_1, r)^\perp$ holds for some $r \in [t_1, t_2]$ then $x(t) = y(t)$ on $[t_1, r]$.
Proof. Define a vector function $(z(t), \nu(t))$ as follows:

$$(z(t), \nu(t)) = (x(t), \eta(t) + (\mu(t_1) - \eta(t_1))\Phi(t_1, t)).$$

The vector function $(z(t), \nu(t))$ is a solution of the system (2.2.1), (2.2.2) over $[t_1, r]$ with $(z(t_1), \nu(t_1)) = (y(t_1), \mu(t_1))$. On the other hand $(y(t), \mu(t))$ has exactly the same property, hence

$$(y(t), \mu(t)) = (x(t), \eta(t) + (\mu(t_1) - \eta(t_1))\Phi(t_1, t)) \text{ on } [t_1, r].$$

This implies that $u(t) = w(t)$ on $[t_1, r]$ and, therefore, $x(t) = y(t)$ for all t which belongs to $[t_1, r]$.

In the next two theorems we establish the connection between the branching of the extremal trajectory $x(t)$ and the points of discontinuity of $dimV(t_1, t)$ as a function of t. In order to derive the next theorem we need to recall the notion of the difference between two sets, as follows:

For each two sets E and F we denote by $E - F$ the difference between E and F, namely, the set of all those points of E which do not belong to F.

<u>Theorem 2.3.1</u> Consider an extremal triple $(y(t), w(t), \mu(t))$. If $y(t)$ branches out of $x(t)$ at r, then
$$\mu(t_1) - \eta(t_1) \in V(t_1, r)^\perp - V(t_1, r^+)^\perp$$
(which means that:

    a) r is a point of discontinuity of $dimV(t_1, t)$,

    b) $V(t_1, r)$ is a proper subspace of $V(t_1, r^+)$).

<u>Proof.</u> For each extremal trajectory $y(t)$ which coincides with $x(t)$ on $[t_1, r]$ the inclusion $\mu(t_1) - \eta(t_1) \in V(t_1, r)^\perp$ holds (due to Lemma 1.4.1). The trajectory $y(t)$ branches out of $x(t)$ at r, this implies that for each $h > 0$ the relation

$$x(r + h) \neq y(r + h)$$

holds. This yields, (due to Lemma 2.3.1) that

$$\mu(t_1) - \eta(t_1) \notin V(t_1, r + h)^\perp.$$

Definition 1.4.2, therefore, implies

$$\mu(t_1) - \eta(t_1) \notin V(t_1, r^+)^\perp.$$

This completes the proof.

<u>Theorem 2.3.2</u>   Assume that r is a point of discontinuity of $dimV(t_1, t)$. Suppose that $(y(t), w(t), \mu(t))$ is an extremal triple such that $x(t_1) = y(t_1)$. If

$$\mu(t_1) - \eta(t_1) \in V(t_1, r)^\perp - V(t_1, r^+)^\perp,$$

then $y(t)$ branches out of $x(t)$ at r.

Proof. Inasmuch as $x(t_1) = y(t_1)$ and $\mu(t_1) - \eta(t_1) \in V(t_1, r)^\perp$ the extremal trajectories $x(t), y(t)$ coincide on the interval $[t_1, r]$ (due to Lemma 2.3.1). On the other hand $\mu(t_1) - \eta(t_1) \notin V(t_1, r^+)^\perp$. Hence, for each $\epsilon > 0$ there exists h such that $\epsilon > h > 0$ and $x(r + h) \neq y(r + h)$. This implies that $x(t)$ and $y(t)$ are different on $[r + h, t_2]$. The arbitrariness of ϵ implies that $x(t) \neq y(t)$ for each $\in (r, t_2]$. This completes the proof.

<u>Definition 2.3.1</u> We denote the set of discontinuity points of the function $dimV(t_1, s)$ on the interval $[t_1, t_2]$ by $\{r_i\}$.

We are now in a position to state and prove the main results of this section, namely the necessary and sufficient conditions for branching.

<u>Theorem 2.3.3</u> Consider a pair (μ, r) such that:

      1. $\mu \in R^n$,

      2. $r \in \{r_i\}$,

      3. $\mu - \eta(t_1) \in V(t_1, r)^\perp - V(t_1, r^+)^\perp$.

There exists an extremal trajectory $y(t; \mu)$ which branches out of $x(t)$ at r. (Here by $y(t; \mu)$ we denote first n coordinates of a solution $(y(t; \mu), \mu(t))$ of $(2.2.1) - (2.2.2)$ with $y(t_1; \mu) = x(t_1)$, $\mu(t_1) = \mu$.)

<u>Proof.</u> An easy consequence of Theorem 2.3.2 .

<u>Theorem 2.3.4</u> Consider an extremal trajectory $y(t)$ such that $y(t_1) = x(t_1)$. If $y(t)$ branches out of $x(t)$ at $r \in [t_1, t_2]$ then:

      1. $r \in \{r_i\}$,

      2. $y(t) = y(t; \mu)$ for some $\mu \in R^n$,

      3. $\mu - \eta(t_1) \in V(t_1, r)^\perp - V(t_1, r^+)^\perp$.

<u>Proof.</u> An easy consequence of Theorem 2.3.1 .

Thus it turns out that the set of branching points does not depend on the chosen extremal trajectory and, therefore, it makes sense to speak about the set of branching points of the system (1.4.1) on the time interval $[t_1, t_2]$. This set is exactly the set of discontinuity points of the function $dimV(t_1, s)$. We denote the set of branching points of the system (1.4.1) on the time interval $[t_1, t_2]$ by $R\{[t_1, t_2]\}$. Note that the relation $t_1 \leq \sigma \leq s$ implies that $R\{[t_1, \sigma]\} \subset R\{[t_1, s]\}$.

§2.4 Special case: linear–quadratic problem

In this section we shall be concerned with the linear control system (1.4.1) with a quadratic cost

$$c(x, u) = \int_{t_1}^{t_2} \begin{pmatrix} x(t) \\ u(t) \end{pmatrix}^* \begin{pmatrix} R_{11}(t) & R_{12}(t) \\ R_{21}(t) & R_{22}(t) \end{pmatrix} \begin{pmatrix} x(t) \\ u(t) \end{pmatrix} dt, \qquad (2.4.1)$$

where $x \in R^n$, $u \in R^m$ and $R_{11}(t)$, $R_{12}(t)$, $R_{21}(t)$, $R_{22}(t)$ are matrices with appropriate dimensionalities whose elements are Lebesgue integrable functions on $[t_1, t_2]$ and $*$ indicates the transpose

For convenience we assume throughout that $R_{12}(t) = R_{21}^*(t)$. We wish to mention, that in this case the cost functional is not necessary convex in (x, u), and the extended attainable set is not necessarily a convex set (consider for example x, u are scalars, the linear control equation is $\dot{x}(t) = u(t)$, and $R_{11}(t) = -1$, $R_{12}(t) = R_{21}(t) = 0$, $R_{22}(t) = 1$).

It turns out that a linearization of the Hamiltonian system of a nonlinear control problem is the Hamiltonian system of a linear control problem of the type (1.4.1), (2.4.1). Namely, in this sense, a linear–quadratic control problem is a first order estimation, or "derivative", of a nonlinear control problem. We will show in Chapter 4, that branching points of a nonlinear control problem are completely determined by it's "derivative". In other words, branching points of an extremal trajectory of a nonlinear control problem are branching points of a linear control problem of the type (1.4.1), (2.4.1). This remark is the main motivation behind the investigation of branching points in the linear–quadratic control problem.

Note that for the linear–quadratic problem Hypothesis 1.3.1 turns into the following:

Hypothesis 2.4.1

1. $R_{22}(t)$ is uniformly positive definite on $[t_1, t_2]$. Namely there exists a positive scalar δ such that $u^* R_{22}(t) u \geq \delta |u|^2$ on $[t_1, t_2]$ for each $u \in R^m$,

2. for each admissible pair $(z(t), v(t))$ with $z(t_1) = z(t_2) = 0$ the cost $c(z, v) \geq 0$ and $c(z, v) = 0$ if and only if $(z(t), v(t)) = (0, 0)$.

In the case of the linear control problem (1.4.1), (2.4.1) Assumption 1.5.1 is fulfiled and one can derive an explicit formula for extremal control as follows:

$$\bar{u}(t, x, \eta) = \frac{1}{2} R_{22}^{*-1}(t)\{B^*(t)\eta^* - 2R_{12}(t)x\}. \tag{2.4.2}$$

The adjoint equation is

$$\dot{\eta}(t) = 2x^*(t)R_{11}(t) + 2u^*(t)R_{12}(t) - \eta(t)A(t) \tag{2.4.3}$$

and the Pontryagin Maximum Principle is the following:

$$-2u^*(t)R_{22}(t) - 2x^*(t)R_{12}(t) + \eta(t)B(t) = 0. \tag{2.4.4}$$

We wish to show now that branching is still a unique possible form of intersection of extremals initiating at the same point. First we mention the following important property of extremals of the linear control problem (1.4.1), (2.4.1):

Theorem 2.4.1 Let $(x(t), u(t))$ be an extremal pair. Then for each admissible pair $(y(t), w(t))$ with $x(t_1) = y(t_1)$ and $x(t_2) = y(t_2)$

$$c(x, u) \leq c(y, w) \text{ and } c(x, u) = c(y, w) \text{ if and only if } (x(t), u(t)) = (y(t), w(t)) \text{ on } [t_1, t_2].$$

Proof. We wish to show that $c(y, w) - c(x, u) \geq 0$. The pair $(x(t), u(t))$ is an extremal pair, hence the first element of the Taylor expansion of the difference vanishes (see e.g. [11], p.357). The second (and the last) term of the expansion is $c(y - x, w - u)$. Due to Hypothesis 2.4.1 $c(y - x, w - u) \geq 0$ and $c(y - x, w - u) = 0$ if and only if $x(t) = y(t)$ and $u(t) = w(t)$. This finishes the proof.

Let $(x(t), u(t))$ and $(y(t), w(t))$ be extremal pairs with $x(t_1) = y(t_1)$ and $x(s) = y(s)$ for some $s \in [t_1, t_2]$. Denote

$$\int_{t_1}^{s} \begin{pmatrix} x(t) \\ u(t) \end{pmatrix}^{*} \begin{pmatrix} R_{11}(t) & R_{12}(t) \\ R_{21}(t) & R_{22}(t) \end{pmatrix} \begin{pmatrix} x(t) \\ u(t) \end{pmatrix} dt \qquad \text{by} \qquad c_s(x, u).$$

Due to Theorem 2.4.1 $c_s(x, u) \leq c_s(y, w)$ and $c_s(y, w) \leq c_s(x, u)$, namely $c_s(y - x, w - u) = 0$ and $y(t) = x(t)$, $w(t) = u(t)$ on $[t_1, s]$. This remark indicates, that branching is a unique possible form of intersection of extremals.

The remainder of the section is devoted to the derivation of necessary and sufficient conditions for branching.

Necessary and sufficient conditions for branching.

Lemma 2.4.1 Let $(x(t), u(t), \eta(t))$ be an extremal triple such that $x(t_1) = 0$ and $\eta(t_1) \in V(t_1, r)^{\perp}$. Then $(x(t), u(t), \eta(t)) = (0, 0, \eta(t_1)\Phi(t_1, t))$ on $[t_1, r]$.

Proof. A straightforward verification shows that the triple $(0, 0, \eta(t_1)\Phi(t_1, t))$ is an extremal triple of the optimal control problem (1.4.1)-(2.4.1) on the initial time interval $[t_1, r]$. Namely, $(0, \eta(t_1)\Phi(t_1, t))$ is a solution of the Hamiltonian system (2.2.1), (2.2.2) on the time interval $[t_1, r]$. Note that $(x(t), \eta(t))$ is also a solution of the same Hamiltonian system. Moreover, these two solutions have the same initial conditions, hence $(x(t), \eta(t)) = (0, \eta(t_1)\Phi(t_1, t))$ on $[t_1, r]$. In accordance with (2.4.2) and the condition $\eta(t_1) \in V(t_1, r)^{\perp}$ the control $u(t)$ is 0 on $[t_1, r]$.

Lemma 2.4.2 Let $(x(t), u(t), \eta(t))$ be an extremal triple such that $x(t) = 0$ on $[t_1, r]$. Then $\eta(t_1) \in V(t_1, r)^{\perp}$.

Proof. Note, that if $x(t) = 0$ on $[t_1, r]$ then

$$0 = \dot{x}(t) = \frac{1}{2}B(t)R_{22}^{*-1}(t)B^{*}(t)\eta^{*}(t) \text{ on } [t_1, r],$$

this implies that

$$0 = \eta(t)B(t)R_{22}^{*-1}(t)B^{*}(t)\eta^{*}(t) \text{ on } [t_1, r].$$

The matrix $R_{22}^{-1}(t)$ is positive definite, hence $\eta(t)B(t) = 0$ on $[t_1, r]$. This implies, due to formula (2.4.2), that $u(t) = 0$ on $[t_1, r]$. Then, due to (2.4.3)

$$\eta(t) = \eta(t_1)\Phi(t_1, t) \text{ on } [t_1, r],$$

and finally equation (2.4.4) yields

$$\eta(t_1)\Phi(t_1,t)B(t) = 0 \text{ on } [t_1,r].$$

In other words $\eta(t_1) \in V(t_1,r)^{\perp}$.

We present now the main result of the section:

<u>Theorem 2.4.2</u> Let $(x(t), u(t), \eta(t))$ and $(y(t), w(t), \mu(t))$ be two extremal triples with $x(t_1) = y(t_1)$. Then $x(t) = y(t)$ on $[t_1,r]$ if and only if $\eta(t_1) - \mu(t_1) \in V(t_1,r)^{\perp}$. If $x(r) = y(r)$, then $(x(t), u(t)) = (y(t), w(t))$ on $[t_1,r]$ and $\eta(t) - \mu(t) = (\eta(t_1) - \mu(t_1))\Phi(t_1,t)$ on $[t_1,r]$.

<u>Proof.</u> Invoke Lemma 2.4.1 and Lemma 2.4.2.

Thus it turns out once again that the set of branching points does not depend on the chosen extremal trajectory and the cost functional. It makes sense, therefore, to speak about the set of branching points of the linear system (1.4.1) on the time interval $[t_1,t_2]$. This set is exactly the set of discontinuity points of the function $dimV(t_1,s)$. In particular in the case where the coefficients of the system are constants, i.e. $A(t) = A$ and $B(t) = B$ there are no branching points at all.

<u>Remark 2.4.1</u> Note that in the case of the linear control system (1.4.1) with the quadratic cost (2.4.1) the Hamiltonian system (2.2.1), (2.2.2) is a system of linear differential equations with respect to (x,η). Let $(x(t;\xi), \eta(t;\xi))$ be a solution of this system with $x(t_1;\xi) = 0$ and $\eta(t_1,\xi) = \xi$. For each time $t \in [t_1,t_2]$ consider a linear mapping $\phi_t : R^n \to R^n$ defined as follows:

$$\phi_t(\xi) = x(t;\xi).$$

In accordance with Theorem 2.4.2 the rank of the matrix $\frac{\partial \phi_t(\cdot)}{\partial \xi}$ is $dimV(t_1,t)$. Namely

$$rank \frac{\partial \phi_t(\cdot)}{\partial \xi} = dimV(t_1,t). \tag{2.4.5}$$

On the other hand $dimV(t_1,t) = dimAT(t;t_1,0)$. Hence for each $x \in AT(t;t_1,0)$ there exists an extremal $x(s)$ with $x(t_1) = 0$ and $x(t) = x$. This remark and the relation (2.4.5) will be extremely useful for future investigation of branching points in nonlinear control problems.

§2.5 Left branching

The third section of this chapter has been devoted mainly to the description of the branching points. In this section we consider left branching points of the control system (1.4.1). Namely, we look for the set $\{l_j\}$ such, that for each $l \in \{l_j\}$ there exist extremal trajectories $x(t)$ and $y(t)$, and the following condition holds:

$$y(t) \neq x(t) \text{ on } [t_1, l) \text{ and } y(t) = x(t) \text{ on } [l, t_2].$$

In this section we follow the introduced terminology and say that $y(t)$ branches out of $x(t)$ at the point l. We call l a left branching point of $x(t)$.

By the same technique as in section 3, we get similar results. The structure of the left branching is absolutely symmetric to the structure of the branching that has been considered in the preceding section. This is why we would like no more than to display the related definitions and to formulate the main theorems without proofs.

<u>Definition 2.5.1</u> We denote the set of discontinuity points of the function $dimV(t_2, s)$ on the interval $[t_1, t_2]$ by $\{l_j\}$.

<u>Theorem 2.5.1</u> Consider a pair (μ, l) such that $\mu \in R^n, l \in \{l_j\}$. If

$$\mu - \eta(t_2) \in V(t_2, l)^{\perp} - V(t_2, l^-)^{\perp},$$

then there exists an extremal trajectory $y(t; \mu)$ which branches out of $x(t)$ at l.

<u>Theorem 2.5.2</u> Suppose that $y(t)$ is an extremal trajectory. If $y(t)$ branches out of $x(t)$ at l then:

        1. $l \in \{l_j\}$,

        2. $y(t) = y(t; \mu)$ for some $\mu \in R^n$,

        3. $\eta(t_2) - \mu \in V(t_2, l)^{\perp} - V(t_2, l^-)^{\perp}$.

We denote the set of left branching points of the system (1.4.1) on the time interval $[t_1, t_2]$ by $L\{[t_1, t_2]\}$. Observe, that $L\{[\sigma, t_2]\} \subset L\{[s, t_2]\}$ provided $s \leq \sigma \leq t_2$.

Chapter 3

Branching pairs in linear control problems

§3.1 Preface

This chapter is devoted to the description of the branching pairs set. Namely, let $x(t)$ be an extremal trajectory defined on a time interval $[t_1, t_2]$. Consider a pair $\{l, r\}$ such that there exists an extremal trajectory $y(t)$ which coincides with $x(t)$ on $[l, r]$, and differs from it on the rest of the time interval $[t_1, t_2]$. We shall say that $y(t)$ branches out of $x(t)$ at $\{l, r\}$, and $\{l, r\}$ is a branching pair of the extremal $x(t)$.

The main result of this chapter is that the set of the branching pairs depends on the matrices $A(t)$, $B(t)$ only. This set, which we denote by $T\{[t_1, t_2]\}$, is countable, in contrast with the finite set of the branching points $R\{[t_1, t_2]\}$.

The structure of this chapter is as follows: We begin with a simple example of a control system with a non–empty set of branching pairs. In the second section a finite family of the sets V_k, $k = 0, 1, .., n - 1$ is introduced. We shall prove that each V_k is at most countable. In the third section we establish the connection between $T\{[t_1, t_2]\}$ and the introduced finite family. Namely, we show that

$$T\{[t_1, t_2]\} = \bigcup_{k=0}^{n-1} V_k.$$

Example 3.1.1 An optimal control problem with a non empty set of branching pairs.

Consider the optimal control problem which has been treated in Example 1.4.1. An extremal trajectory $x(t)$ has the following form:

$$x(t) = x(0) + \gamma(t)$$

where the vector $\gamma(t)$ is given by:

$$\begin{pmatrix} t\eta_1 \\ 0 \end{pmatrix} \text{ for } 0 \le t < \frac{1}{3}, \quad \begin{pmatrix} \frac{1}{3}\eta_1 \\ 0 \end{pmatrix} \text{ for } \frac{1}{3} \le t < \frac{2}{3}, \quad \begin{pmatrix} \frac{1}{3}\eta_1 \\ (t - \frac{2}{3})\eta_2 \end{pmatrix} \text{ for } \frac{2}{3} \le t \le 1.$$

Here η_1, η_2 are the components of the vector η. Choose x_0, y_0 and the corresponding vectors $\eta = (\eta_1, \eta_2)$, $\mu = (\mu_1, \mu_2)$ as follows:

$$\eta_1 \ne \mu_1, \quad \eta_2 \ne \mu_2, \quad y_0 = x_0 + \begin{pmatrix} \frac{1}{3}(\eta_1 - \mu_1) \\ 0 \end{pmatrix}.$$

Let $(x(t), \eta(t))$ and $(y(t), \mu(t))$ be solutions of the Hamiltonian system (2.2.1), (2.2.2) with $(x(0), \eta(0)) = (x_0, \eta)$ and $(y(0), \mu(0)) = (y_0, \mu)$. This choice of the initial conditions implies that $x(t) = y(t)$ for $\frac{1}{3} \le t \le \frac{2}{3}$, and $x(t) \ne y(t)$ otherwise. This shows, that $\{\frac{1}{3}, \frac{2}{3}\}$ is a branching pair of the system (2.1).

§3.2 Preliminaries

In this section we define the finite collection of disjoint sets V_k, $k = 0, 1, .., n-1$. Our goal is the description of this sets. As will soon be shown (Corollary 3.3.1)

$$T\{[t_1, t_2]\} = \bigcup_{k=0}^{n-1} V_k.$$

At the end of the section (Lemma 3.2.2) it turns out that each V_k is at most a countable set.

<u>Definition 3.2.1</u> For each integer k, $0 \le k \le n-1$ we denote by V_k the set of all pairs $\{l, r\}$ such that:

1. $t_1 < l < r < t_2$,
2. $dimV(l, r) = k$,
3. $dimV(l, r^+) > k$,
4. $dimV(r, l^-) > k$.

Observe that for two different pairs $\{l, r\}$ and $\{l_1, r_1\}$ which belong to the same V_k the relation

$$[l_1, r_1] \subset [l, r] \text{ and } [l, r] \subset [l_1, r_1]$$

is impossible.

Our aim is to prove that each V_k is at most a countable set. The main idea of the proof is as follows: Consider two different pairs $\{l, r\}$, $\{l_1, r_1\}$ which belong to the same V_k. The previous observation implies that

$$r > r_1 \text{ and } l > l_1 \quad \text{or} \quad r_1 > r \text{ and } l_1 > l.$$

Suppose, for a moment, that $r_1 > r$ and $l_1 > l$, namely, the half open subinterval $[l, l_1)$ does no intersect $[l_1, r_1]$. We wish to find such a point $LB(l, r) \in [l, r]$, that for each $\{l^*, r^*\} \in V_k$ such that $r^* > r$ and $l^* > l$ the relation

$$[l, LB(l, r)) \cap [l^*, r^*] = \emptyset \text{ holds.}$$

In order to prove the existence of the point $LB(l,r)$, some preliminary definitions and lemmas are needed.

<u>Definition 3.2.2</u>

$$LB(l,r) = \sup\{\ t \mid t \leq r,\ V(r,t^-) = V(r,l)\ \}.$$

Note that $LB(l,r)$ is always greater than l and $dimV(r,LB(l,r)) < dimV(r,l)$.

Let Φ be a $n \times n$ matrix and V a subspace of R^n. Denote by ΦV, $V\Phi$ the following sets:

$$\Phi V = \{w \in R^n \mid w = \Phi v \text{ for some } v \in V\},$$

$$V\Phi = \{w \in R^n \mid w = v\Phi \text{ for some } v \in V\}.$$

Roughly speaking, Lemma 3.2.1 below asserts the following: For each two pairs $\{l,r\}$ and $\{l_1,r_1\}$ which belong to the same V_k the intervals $[l_1,r_1]$ and $[l, LB(l,r))$ do not overlap.

<u>Lemma 3.2.1</u> Assume that $\{l,r\}$ and $\{l_1,r_1\}$ belong to V_k. If $r_1 > r$ then $l_1 \geq LB(l,r)$.

The idea of the proof is much simpler than the proof itself. We assume the opposite, namely

$$r_1 > r \text{ and } l_1 < LB(l,r).$$

First note that $[a,b] \subset [c,d]$ ensures that

$$dimV(c,d) \geq dimV(a,b).$$

The assumption implies that

$$dimV(r_1,l_1) \geq dimV(r,l_1).$$

Note that $dimV(l,r^+) > k$. This means that there is a "jump of dimension" at r. This implies the strong inequality, namely

$$dimV(r_1,l_1) > dimV(r,l_1).$$

On the other hand

$$k = dimV(r,l) \geq dimV(r,l_1) \geq dimV(r,LB(l,r)^-) = k.$$

This yields

$$dimV(r,l_1) = k.$$

Finally we obtain

$$k = dimV(r_1,l_1) > dimV(r,l_1) = k$$

–a contradiction.

More precisely:

Proof. Suppose the opposite, namely, that $l_1 < LB(l,r)$. Using the definitions we obtain the following relation:

$$V(l,r_1) = V(l,l_1) + \Phi(l,l_1)V(l_1,r) + \Phi(l,r)V(r,r_1).$$

Definition 3.2.2 implies that

$$V(l,l_1) + \Phi(l,l_1)V(l_1,r) = \Phi(l,l_1)V(l_1,r).$$

This yields

$$V(l,r_1) = \Phi(l,l_1)V(l_1,r) + \Phi(l,r)V(r,r_1).$$

On the other hand

$$V(l_1,r_1) = V(l_1,r) + \Phi(l_1,r)V(r,r_1) = \Phi(l_1,l)\{\Phi(l,l_1)V(l_1,r) + \Phi(l,r)V(r,r_1)\}.$$

It then follows that

$$V(l_1,r_1) = \Phi(l_1,l)V(l,r_1),$$

and finally

$$k = dimV(l_1,r_1) = dim\Phi(l_1,l)V(l,r_1) = dimV(l,r_1) \geq dimV(l,r^+) > k.$$

This contradiction completes the proof.

We are now able to state and prove the main result of this section.

Lemma 3.2.2 For each integer k the set V_k is at most countable.

The idea of a proof is as follows: We associate each pair $\{l,r\} \in V_k$ with a certain rational number $rn(l,r)$ which belongs to $[l, LB(l,r))$. By using Lemma 3.2.1 we prove that different elements of V_k are associated with different rationals.

Proof. For each $\{l,r\} \in V_k$ choose a rational number $rn(l,r)$ such that $rn(l,r)$ belongs to $(l, LB(l,r))$.

Claim. If $[l,r] \neq [l_1,r_1]$ then $rn(l,r) \neq rn(l_1,r_1)$.

Proof. If $[l,r] \neq [l_1,r_1]$ then either $r > r_1$ and $l > l_1$ or $r_1 > r$ and $l_1 > l$. If $r > r_1$ and $l > l_1$ then $rn(l,r) > l > LB(l_1,r_1) > rn(l_1,r_1)$. If $r_1 > r$ and $l_1 > l$ then $rn(l_1,r_1) > l_1 > LB(l,r) > rn(l,r)$ This completes the proofs of the claim and the lemma.

§3.3 Decomposition of the set of branching pairs

In this section our main purpose is to show that

$$T\{[t_1, t_2]\} = \bigcup_{k=0}^{n-1} V_k.$$

We begin with some necessary auxiliary statements.

Consider a pair $\{l, r\} \in V_k$. Lemma 1.4.1 implies that for each extremal trajectory $y(t)$ which coincides with $x(t)$ on $[l, r]$

$$\eta(l) - \mu(l) \in V(l, r)^{\perp}.$$

(Here $(y(t), \mu(t))$ is a solution of (2.2.1)–(2.2.2) generated by the extremal trajectory $y(t)$, see Chapter 2). On the other hand (due to Lemma 1.4.1)

$$\eta(r) - \mu(r) \in V(r, l)^{\perp} \text{ and } \eta(r) - \mu(r) = (\eta(l) - \mu(l))\Phi(l, r),$$

hence $V(r, l)^{\perp} = V(l, r)^{\perp} \Phi(l, r)$.

We, therefore, get the following result:

Lemma 3.3.1

$$V(r, l)^{\perp} - \{ V(r, l^-)^{\perp} \cup V(l, r^+)^{\perp} \Phi(l, r) \} \neq \emptyset.$$

Proof. If $\{l, r\} \in V_k$, then $V(r, l)$ is a proper subspace of $V(r, l^-)$. This implies that $V(r, l^-)^{\perp}$ is a proper subspace of $V(r, l)^{\perp}$. On the other hand $V(l, r^+)^{\perp}$ is a proper subspace of $V(l, r)^{\perp}$ and, therefore, $V(l, r^+)^{\perp} \Phi(l, r)$ is a proper subspace of $V(r, l)^{\perp}$.

In the next two lemmas we establish a one-to-one correspondence between the set of the branching pairs of $x(t)$ and the set $\bigcup_{k=0}^{n-1} V_k$.

Lemma 3.3.2 Consider an extremal trajectory $y(t)$ and a pair $\{l, r\}$. If $x(t) = y(t)$ over $[l, r]$ and $x(t) \neq y(t)$ otherwise, then $\{l, r\} \in V_k$ for some $0 \leq k \leq n - 1$.
Proof. Look at $\eta_l = \eta(l) - \mu(l)$ and $\eta_r = \eta(r) - \mu(r)$. Clearly η_l is different from 0 and $\eta_l \in V(l, r)^{\perp} - V(l, r^+)^{\perp}$. Hence, $V(l, r)$ is a proper subspace of $V(l, r^+)$. On the other hand $\eta_r = \eta_l \Phi(l, r)$. This implies that

$$0 \neq \eta_r \in V(r, l)^{\perp} - V(r, l^-)^{\perp}$$

and, therefore, $V(r, l)$ is a proper subspace of $V(r, l^-)$. This means that

$$\{l, r\} \in V_k \text{ for } k = dimV(l, r) = dimV(r, l).$$

Lemma 3.3.3 Assume that $\{l, r\} \in V_k$ for some k. Then there exists an extremal trajectory $y(t)$ which branches out of $x(t)$ at $\{l, r\}$.

Proof. Choose $\eta_r \neq 0$ such that

$$\eta_r \in V(r,l)^\perp - \{V(r,l^-)^\perp \cup V(l,r^+)^\perp \Phi(l,r)\}$$

(such a choice is possible due to Lemma 3.3.1). Let $(y(t), w(t))$ be a solution of the Hamiltonian system (1.5.1), (1.5.2) with

$$(y(r), \mu(r)) = (x(r), \eta(r) + \eta_r).$$

Then the trajectory $y(t)$ is the desired extremal trajectory.

Now, by using the above results it is possible to derive the following statements:

Corollary 3.3.1 $\bigcup_{k=0}^{n-1} V_k$ is the set of two sided branching pairs.
Proof. An easy consequence of Lemma 3.3.2 and Lemma 3.3.3.

Remark 3.3.1 The linear control system (1.4.1), (1.3.2) with constant coefficients, i.e. $A(t) = A$, $B(t) = B$, has no branching at all.

Corollary 3.3.2 $T\{[t_1, t_2]\} \supset T\{[s_1, s_2]\}$ provided $[t_1, t_2] \supset [s_1, s_2]$.

Theorem 3.3.1 $T\{[t_1, t_2]\}$ is at most a countable set.
Proof. For each k such that $0 \leq k \leq n - 1$, V_k is at most countable. The rest of the statement follows from Corollary 3.3.1.

We wish to emphasize that Theorem 3.3.1 is the best possible result. Namely, there exist linear control systems with countable sets of branching pairs (see Example 1.8.1).

We present now the main result of the chapter:
Theorem 3.3.2 A pair $\{l, r\}$ is a branching pair if and only if the following condition satisfied:

$V(l, r)$ is a proper subspace of $V(l, r^+)$ and $V(r, l)$ is a proper subspace of $V(r, l^-)$.

Proof. The proof follows from Lemma 3.3.2 and Lemma 3.3.3.

Remark 3.3.2 The results of this section remain true also for the linear–quadratic problem (1.4.1), (2.4.1).

Chapter 4

The nonlinear case

§4.1 Preliminary discussion

In this chapter we investigate branching points of an extremal trajectory $x(t)$. The set of branching points of the extremal trajectory $x(t)$ is the set of all points $r \in (t_1, t_2)$ such that, there exists a neighboring extremal trajectory $y(t)$ and the following relation holds:

$$x(t) = y(t) \text{ on } [t_1, r] \text{ and } x(t) \neq y(t) \text{ on } (r, t_2].$$

By an extremal trajectory we mean a trajectory which satisfies the Pontryagin Maximum Principle. We wish to recall that an extremal trajectory $y(t)$ is a neighboring extremal trajectory if the norm $|(x, \eta) - (y, \mu)|_{C[t_1, t_2]} < \epsilon^*$ (see Definition 1.3.5). The value of the ϵ^* will be determined later in the chapter (see Definition 4.3.1). The main result of the chapter is the following: *Let $(x(t), u(t), \eta(t))$ be an extremal triple such that Hypothesis 1.3.1 and Assumption 1.5.1 are satisfied. The set of branching points of $x(t)$ coinsides with that of the linear control system*

$$\dot{z}(t) = A(t)z(t) + B(t)v(t)$$

where $A(t) = \frac{\partial F}{\partial x}(t, x(t), u(t))$ and $B(t) = \frac{\partial F}{\partial u}(t, x(t), u(t))$.

However, in order to obtain the result we need to impose an additional restriction. Consider a solution $(y(t), \mu(t))$ of the Hamiltonian system (1.5.1), (1.5.2) with $y(t_1) = x_1$, $\mu(t_1) = \xi$. Denote this solution by $(x(t; x_1, \xi), \eta(t; x_1, \xi))$. Once selected, $x(t)$ will not be changed throughout this section. Hence, this solution can be unambiguously denoted by $(x(t; \xi), \eta(t; \xi))$.

We wish to ensure the smooth (C^1) dependence of solutions $\{x(t; \xi)\}$ on initial conditions ξ. To this end an additional assumption is introduced.

Assumption 4.1.1 There exists a Lebesgue integrable function $m(t)$ such that for each extremal triple $(y(t), w(t), \mu(t))$ generated by $\mu(t_1)$ closed to $\eta(t_1)$ the following condition holds:

$$\left| \begin{pmatrix} H_\eta(t, y(t), w(t), \mu(t)) \\ -H_x(t, y(t), w(t), \mu(t)) \end{pmatrix} \right| + \left| \begin{pmatrix} H_{\eta x}(t, y(t), w(t), \mu(t)), & H_{\eta \eta}(t, y(t), w(t), \mu(t)) \\ -H_{xx}(t, y(t), w(t), \mu(t)), & -H_{x\eta}(t, y(t), w(t), \mu(t)) \end{pmatrix} \right| \leq m(t).$$

The mapping $x(t;\cdot) : R^n \mapsto R^n$ ($\xi \mapsto x(t;\xi)$) is, therefore, continuously differentiable with respect to ξ. (A proof can be easily constructed with minor changes from that of Theorem 7.2, p. 25 in [4].)

We wish now to discuss the relation between the following two conditions:

$$\max_u H(t,x,u,\eta) = H(t,x,u(t,x,\eta),\eta) \ \text{and} \ H_u(t,x,u(t,x,\eta),\eta) = 0.$$

We call the first one the Pontryagin condition and the second one the Euler condition. A function $u(t,x,\eta)$ which satisfies the Pontryagin (respectively Euler) condition is called a Pontryagin (respectively Euler) control function.

It is clear that every Pontryagin control function is simultaneously an Euler control function, but the converse is in general false (see Example 1.10.2). Due to Assumption 1.5.1 the control function $u(t,x,\eta)$ considered in this work is a Pontryagin control function.

The Euler condition allows to express a control function as a function of (t,x,η) by using the Implicit Function Theorem (see Spivak [21], p.41). Such a representation will be extremely useful in our study. Fortunately one can prove that under Assumption 1.5.1 and Hypothesis 1.3.1 there is an $\epsilon > 0$ such that for each pair (y,μ) satisfies:

$$|y - x(t)| + |\mu - \eta(t)| < \epsilon$$

these conditions are equivalent, namely

$$\max_u H(t,y,u,\mu) = H(t,y,v,\mu) \ \text{if and and only if} \ H_u(t,y,v,\mu) = 0.$$

Indeed, the Hypothesis 1.3.1 implies that $det\ H_{uu}(t,x(t),u(t),\eta(t)) \neq 0$. By the Implicit Function Theorem (see Spivak, [21], p. 41) there is $\epsilon(t) > 0$ such that for each (y,μ) with $|y - x(t)| + |\mu - \eta(t)| < \epsilon(t)$ there is a unique $v(t,y,\mu)$ such that $H_u(t,y,v(t,y,\mu),\mu) = 0$ holds on $[t_1,t_2]$. Denote $\inf_{t \in [t_1,t_2]} \epsilon(t)$ by ϵ_1. Note, that due to the first condition of Hypothesis 1.3.1, this ϵ_1 is a positive constant. Suppose now that $|y - x(t)| + |\mu - \eta(t)| < \epsilon_1$. This implies that $H_u(t,y,v(t,y,\mu),\mu) = 0$. On the other hand

$$H(t,y,u(t,y,\mu),\mu) = \max_u H(t,y,u,\mu),$$

hence

$$H_u(t,y,u(t,y,\mu),\mu) = 0.$$

Uniqueness of $v(t,y,\mu)$ (due to the Implicit Function Theorem) implies that

$$v(t,y,\mu) = u(t,y,\mu),$$

namely the condition $H_u(t,y,v,\mu) = 0$ is the necessary and sufficient condition for maximum.

The results of this Chapter **do not** depend on a particular choice of a function $u(t, x, \eta)$. Namely, one can substitute Assumption 1.5.1 by the weaker Assumption 1.5.1':

<u>Assumption 1.5.1'</u> For each triple (t, x, η) close to $(t, x(t), \eta(t))$ there exists a unique $u(t, x, \eta)$ continuously differentiable with respect to (x, η) such that the relation

$$H_u(t, x, u(t, x, \eta), \eta) = 0 \text{ holds on } [t_1, t_2].$$

§4.2 Matching of extremals

Throughout this section we examine the extremals which initiate at the same point $x(t_1)$ as $x(t)$ and the distance $|\eta(t_1) - \mu(t_1)|$ is small. The necessary conditions for matching of these extremals on an initial subinterval $[t_1, r]$ will be derived. Namely, we shall derive the conditions which must be satisfied by each neighboring extremal $y(t)$ which coincides with $x(t)$ on an initial subinterval $[t_1, r]$.

Consider an extremal triple $(y(t), w(t), \mu(t))$, different from $(x(t), u(t), \eta(t))$ such that the relation $x(t_1) = y(t_1)$ holds. We suppose that $w(t) = u(t)$ over an initial subinterval $[t_1, r]$ of the time interval $[t_1, t_2]$. Denote $\mu(t) - \eta(t)$ by $\Delta(t)$. Equation 1.3.3 yields the following:

$$\dot{\Delta}(t) = -\Delta(t)A(t) \quad t \in [t_1, r].$$

As is customary the symbol $\Phi(t, t_1)$ denotes the fundamental solution of the linear differential equation

$$\dot{x} = A(t)x \text{ with } \Phi(t_1, t_1) = I, \text{ where } A(t) = \frac{\partial F}{\partial x}(t, x(t), u(t)).$$

This implies that

$$\Delta(t) = \Delta(t_1)\Phi(t, t_1), \quad \Delta(t_1) = \mu(t_1) - \eta(t_1).$$

On the other hand using equation (1.5.4) and Assumption 1.5.1 we obtain that

$$\Delta(t)B(t) = 0 \text{ on } [t_1, r], \text{ where } B(t) = \frac{\partial F}{\partial u}(t, x(t), u(t)).$$

In other words $\Delta(t_1)\Phi(t_1, t)B(t) = 0$ on $[t_1, r]$. \hfill (4.2.1)

Finally, we obtain the following result:

Lemma 4.2.1 If $(y(t), w(t), \mu(t))$ is an extremal triple such that $w(t) = u(t)$ on $[t_1, r]$ for some $\in (t_1, t_2)$, then $\mu(t_1) - \eta(t_1) \in V(t_1, r)^{\perp}$.

In the next lemma we establish conditions which ensure matching of extremal controls.

<u>Lemma 4.2.2</u> Consider an extremal triple $(y(t), w(t), \mu(t))$ such that

$$y(t_1) = x(t_1) \text{ and } |\mu(t_1) - \eta(t_1)| \text{ is small.}$$

If $\mu(t_1) - \eta(t_1) \in V(t_1, r)^{\perp}$, then $w(t) = u(t)$ on $[t_1, r]$.

($|\mu(t_1) - \eta(t_1)|$ is small means that $|(\mu(t_1) - \eta(t_1))\Phi(t_1, t)| < \epsilon_1$ for each $t \in [t_1, t_2]$.)

Denote $(\mu(t_1) - \eta(t_1))\Phi(t_1, t)$ by $\Delta(t)$. The proof consists of the following three steps:

First we show that the triple $(x(t), u(t), \eta(t) + \Delta(t))$ is an extremal triple of the control problem (1.3.1), (1.3.2) on $[t_1, r]$.

Secondly, note that $(x(t), \eta(t) + \Delta(t))$ and $(y(t), \mu(t))$ are the solutions of the Hamiltonian system (1.5.1), (1.5.2) over an initial subinterval $[t_1, r]$ with the same initial conditions $(y(t_1), \mu(t_1))$. This yields

$$x(t) = y(t), \quad \eta(t) + \Delta(t) = \mu(t) \text{ on } [t_1, r].$$

Thirdly, using the results of the the first and second steps we point out the following relation:

$$u(t) = u(t, x(t), \eta(t)) = u(t, x(t), \eta(t) + \Delta(t)) = u(t, y(t), \mu(t)) = w(t) \text{ on } [t_1, r].$$

This will finish the proof.

<u>Proof.</u> We wish to show that that the triple $(x(t), u(t), \eta(t) + \Delta(t))$ is an extremal triple of the control problem (1.3.1), (1.3.2) over $[t_1, r]$. Note that the relation

$$H(t, x(t), u(t), \eta(t) + \Delta(t)) = \max_u H(t, x(t), u, \eta(t) + \Delta(t)) \text{ holds on } [t_1, r].$$

Indeed, due to the definition of $u(t, x, \eta)$

$$H_u(t, x(t), u(t), \eta(t)) = 0 \text{ on } [t_1, t_2],$$

and $H_u(t, x(t), u(t), \eta(t) + \Delta(t)) = H_u(t, x(t), u(t), \eta(t)) + \Delta(t_1)\Phi(t_1, t)B(t)$ on $[t_1, r]$.

The definition of $\Delta(t_1)$ implies that

$$\Delta(t_1)\Phi(t_1, t)B(t) = 0 \text{ on } [t_1, r].$$

It then follows that

$$H_u(t, x(t), u(t), \eta(t) + \Delta(t)) = 0 \text{ on } [t_1, r].$$

Suppose that

$$\max_u H(t, x(t), u, \eta(t) + \Delta(t)) = H(t, x(t), v(t), \eta(t) + \Delta(t)) \text{ on } [t_1, r].$$

The last relation yields

$$H_u(t, x(t), v(t), \eta(t) + \Delta(t)) = 0 \text{ on } [t_1, r],$$

this implies (due to the Implicit Function Theorem, see [21], p. 41) that $u(t) = v(t)$ on $[t_1, r]$, namely $u(t) = u(t, x(t), \eta(t) + \Delta(t))$ and, therefore, $(x(t), u(t), \eta(t) + \Delta(t))$ is indeed an extremal triple. This completes the proof.

The next theorem shows that the condition $dim\ V(t_1, t) = constant$ is sufficient for non existence of branching points.

<u>Theorem 4.2.1</u> Suppose that $V(t_1, r) = V(t_1, r + h)$ for some $h > 0$. Consider an extremal triple $(y(t), w(t), \mu(t))$ such that

$$x(t_1) = y(t_1) \text{ and } |\eta(t_1) - \mu(t_1)| \text{ is small .}$$

If $w(t) = u(t)$ on $[t_1, r]$, then $w(t) = u(t)$ on $[t_1, r + h]$.

<u>Proof.</u> Due to Lemma 4.2.1 $\mu(t_1) - \eta(t_1) \in V(t_1, r)^{\perp}$. The condition $V(t_1, r) = V(t_1, r + h)$ implies that

$$V(t_1, r)^{\perp} = V(t_1, r + h)^{\perp} \text{ and, therefore, } \mu(t_1) - \eta(t_1) \in V(t_1, r + h)^{\perp}.$$

We now employ Lemma 4.2.2, and this completes the proof.

<u>Corollary 4.2.1</u> If $dim\ V(t_1, r) = n$ for some $r \in (t_1, t_2)$, and $(y(t), w(t), \mu(t))$ is an extremal triple such that $x(t_1) = y(t_1)$ and $u(t) = w(t)$ on $[t_1, r]$, then $u(t) = w(t)$ on $[t_1, t_2]$.

As a matter of fact we wish to show that the matching of extremal trajectories $y(t)$ and $x(t)$ implies matching of the corresponding controls $w(t)$ and $u(t)$. In order to achieve this goal we need an additional condition, namely we require that

$$\mu(t_1) - \eta(t_1) \in V(t_1, r)^{\perp}.$$

Our aim is to show that the behavior of the extremals closed to $x(t)$ is "similar" to the behavior of the extremals in a linear control system with quadratic cost. From this point of view the imposed additional condition looks somewhat artificial. Indeed, this condition will be removed at the end of the next section (Theorem 4.3.1).

The next two statements show that the matching of extremal trajectories is, roughly speaking, the same as the matching of extremal controls.

<u>Lemma 4.2.3</u> Consider an extremal triple $(y(t), w(t), \mu(t))$ such that $|\mu(t_1) - \eta(t_1)|$ is small. Suppose that on some subinterval $[t_1, r]$ of $[t_1, t_2]$ the following relations hold:

$$y(t) = x(t), \quad \mu(t_1) - \eta(t_1) \in V(t_1, r)^{\perp}.$$

Then $w(t) = u(t)$ on $[t_1, r]$.

Proof. Note that the condition $x(t) = y(t)$ implies, in particular, that $y(t_1) = x(t_1)$ and then Lemma 4.2.3 is a simple consequence of Lemma 4.2.2.

Theorem 4.2.2 Consider an extremal triple $(y(t), w(t), \mu(t))$ such that $x(t_1) = y(t_1)$. Suppose that the following conditions hold:

$$\mu(t_1) - \eta(t_1) \in V(t_1, r)^{\perp}, \quad |\mu(t_1) - \eta(t_1)| \text{ is small.}$$

Then $w(t) = u(t)$ on $[t_1, r]$ if and only if $y(t) = x(t)$ on $[t_1, r]$.

Proof. The "only if" part is trivial, the "if" part follows from Lemma 4.2.3.

One of the tasks ahead is to remove the condition

$$\mu(t_1) - \eta(t_1) \in V(t_1, r)^{\perp}$$

imposed in Lemma 4.2.3 and Theorem 4.2.2. This will be done in Theorem 4.3.1.

§4.3 Branching of extremals

The results obtained so far concern the existence of branching points in the intervals of the form $[r, r+h]$ under the condition $V(t_1, r) = V(t_1, r+h)$. Loosely speaking, the central result of the preceding section implies the absence of the branching points in this case. In this section we investigate the branching points of $x(t)$ on the time interval $[r, r+h]$ under the condition

$$dim V(t_1, r) < dim V(t_1, r+h).$$

Consider a solution $(x(t), \eta(t))$ of the Hamiltonian system (1.5.1), (1.5.2) with $x(t_1) = x_1$, $\eta(t_1) = \xi$. Denote this solution by $(x(t; \xi), \eta(t; \xi))$. The questions of considerable interest are "how many different ξ steer x_1 to the same terminal point $x(t; \eta(t_1))$?", or, to put it another way, "how many solutions does an equation $x(t; \xi) = x(t; \eta(t_1))$ have?". In other words "how many neighboring extremals intersect $x(s; \eta(t_1))$ at time t?" The answers to these questions are closely related to knowledge of $rank \left. \frac{\partial x(t; \cdot)}{\partial \xi} \right|_{\eta(t_1)}$, where by $\left. \frac{\partial x(t; \cdot)}{\partial \xi} \right|_{\eta(t_1)}$ we denote the partial derivative of $x(t; \xi)$ with respect to the second argument evaluated at the point $(t; \eta(t_1))$.

In order to determine the $rank \left. \frac{\partial x(t; \cdot)}{\partial \xi} \right|_{\eta(t_1)}$ we consider the auxiliary linear control system

$$\dot{z}(t) = A(t)z(t) + B(t)v(t), \quad z(t_1) = 0. \tag{4.3.1}$$

with a cost functional

$$\int_{t_1}^{t_2} h(t, z(t), v(t)) dt \tag{4.3.2}$$

where $h(t, z, v)$ is a quadratic function with respect to (z, v), namely:

$$h(t, z, v) = \begin{pmatrix} z \\ v \end{pmatrix}^* \begin{pmatrix} R_{11}(t) & R_{12}(t) \\ R_{21}(t) & R_{22}(t) \end{pmatrix} \begin{pmatrix} z \\ v \end{pmatrix},$$

such that Hypothesis 2.4.1 is satisfyed. In this case the adjoint equation has the following form:

$$\dot{\theta} = \frac{\partial h}{\partial z}(t, z(t), v(t)) - \theta A(t)$$

For the linear–quadratic control problem (4.3.1), (4.3.2) $rank \frac{\partial z(t; \cdot)}{\partial \theta}|_0$ is the dimension of the attainable set at time t. Specifically

$$rank \frac{\partial z(t; \cdot)}{\partial \theta}|_0 = dim\, V(t_1, t) \tag{4.3.3}$$

(See Remark 2.4.1, equation 2.4.5).

We intend to construct the linear–quadratic control problem (4.3.1) – (4.3.2) in such a way that

$$rank \frac{\partial x(t; \cdot)}{\partial \xi}|_{\eta(t_1)} = rank \frac{\partial z(t; \cdot)}{\partial \theta}|_0 \text{ for each } t \in [t_1, t_2],$$

where $x(t; \eta(t_1))$ is the extremal trajectory of the control problem (1.3.1), (1.3.2), and $z(t; 0)$ is an extremal trajectory of the linear–quadratic control problem (4.3.1) – (4.3.2). The matrices $A(t)$ and $B(t)$ have been already defined and all that we have left to construct is a cost functional $h(t, z, v)$. Define $h(t, z, v)$ as follows:

$$h(t, z, v) = -\frac{1}{2} \begin{pmatrix} z \\ v \end{pmatrix}^* \begin{pmatrix} H_{xx}(t, x(t), u(t), \eta(t)) & H_{ux}(t, x(t), u(t), \eta(t)) \\ H_{xu}(t, x(t), u(t), \eta(t)) & H_{uu}(t, x(t), u(t), \eta(t)) \end{pmatrix} \begin{pmatrix} z \\ v \end{pmatrix}.$$

Here by * we indicate the transpose. Note that Hypothesis 1.3.1 implies the fulfillment of Hypothesis 2.4.1 for the linear–quadratic problem (4.3.1), (4.3.2).

Denote by the $(n + n) \times (n + n)$ matrix

$$\Psi(t, t_1) = \begin{pmatrix} \Psi_{11}(t, t_1) & \Psi_{12}(t, t_1) \\ \Psi_{21}(t, t_1) & \Psi_{22}(t, t_1) \end{pmatrix} \text{ with } \Psi(t_1, t_1) = I$$

a fundamental solution of the following linear differential equation:

$$\dot{X} = \begin{pmatrix} H_{\eta x}(t, x(t), u(t), \eta(t)) & H_{\eta \eta}(t, x(t), u(t), \eta(t)) \\ -H_{xx}(t, x(t), u(t), \eta(t)) & -H_{x\eta}(t, x(t), u(t), \eta(t)) \end{pmatrix} X,$$

where $X \in R^{n+n}$. In this case

$$\frac{\partial x(t; \cdot)}{\partial \xi}|_{\eta(t_1)} = \Psi_{12}(t, t_1) \tag{4.3.4}$$

see Coddington and Levinson [3], p. 25). On the other hand a direct computation shows that

$$\frac{\partial z(t; \cdot)}{\partial \theta}|_0 = \Psi_{12}(t, t_1) \text{ and, therefore, } \frac{\partial z(t; \cdot)}{\partial \theta}|_0 = \frac{\partial x(t; \cdot)}{\partial \xi}|_{\eta(t_1)}.$$

This implies that $rank \dfrac{\partial x(t;\cdot)}{\partial \xi}|_{\eta(t_1)} = dim\ V(t_1,t)$ (see 4.3.3).

Hence, due to the Implicit Function Theorem, for each $t \in [t_1,t_2]$ there exists a constant $\epsilon(t) > 0$, such that the set

$$\Xi(t) = \{\ \xi\ |\ \xi \in R^n,\ |\xi - \eta(t_1)| < \epsilon(t),\ x(t;\xi) = x(t;\eta(t_1))\ \}$$

is a subset of a $n\text{-}dim V(t_1,t)$ dimensional manifold.

In what follows we describe the family of the sets $\Xi(t)$. In particular we show, that $\Xi(t)$ itself is a $n\text{-}dim V(t_1,t)$ dimensional manifold for each $t \in [t_1,t_2]$. First, choose a positive ϵ_2 such that the condition $|\Delta\Phi(t_1,s)| < \epsilon(s)$ holds for each $|\Delta| < \epsilon_2,\ t_1 \le s \le t_2$. Next, choose a vector Δ in such a way that:

1. $|\Delta| < \epsilon_2$,
2. $\Delta \in V(t_1,t)^\perp$.

(The continuity of $\frac{\partial x(t;\cdot)}{\partial \xi}|_{\eta(t_1)}$ in t (see equation (4.3.4)) implies existence of such positive ϵ_2.) For each vector Δ satisfying conditions 1–2 consider an extremal triple

$$(y(s), w(s), \eta(s) + \Delta\Phi(t_1,s)).$$

Due to Lemma 4.2.2 the corresponding extremal pair $(y(s), w(s))$ coincides with $(x(s), u(s))$ on $[t_1,t]$, in particular $y(t) = x(t)$. The collection of vectors Δ which satisfy conditions 1–2 above is a linear manifold. The dimension of this linear manifold is

$$dim\ V(t_1,t)^\perp = n - dim\ V(t_1,t).$$

This implies that

$$\Xi(t) = \{\ \eta(t_1) + \Delta\ |\ \Delta\ \text{satisfies conditions 1-2}\ \}.$$

and, fortunately, $\Xi(t)$ itself is a $n\text{-}dim V(t_1,t)$ dimensional linear manifold.

We therefore get:

Lemma 4.3.1
$$\Xi(t) = \{\ \xi\ |\ |\xi - \eta(t_1)| < \epsilon_2,\ \ \xi - \eta(t_1) \in V(t_1,t)^\perp\ \}.$$

Corollary 4.3.1
$$\Xi(s) \subset \Xi(\sigma)\ \text{provided}\ s \ge \sigma.$$

Proof. If $s \ge \sigma$ then
$$V(t_1,\sigma) \subset V(t_1,s);$$

this implies

$$V(t_1, s)^\perp \subset V(t_1, \sigma)^\perp.$$

The result now follows from Lemma 4.3.1 .

At this point we are able to present the definition of the positive scalar ϵ^*, which defines the family of the neighboring extremals, as follows:

<u>Definition 4.3.1</u> Denote $\min(\epsilon_1, \epsilon_2)$ by ϵ^*. (We wish to remind the reader that ϵ_1 was defined in section 4.1.) An extremal trajectory $y(t)$ is a neighboring extremal trajectory if

$$\max_{t_1 \leq t \leq t_2} |x(t) - y(t)| + \max_{t_1 \leq t \leq t_2} |\eta(t) - \mu(t)| < \epsilon^*.$$

We can now remove the condition $\mu(t_1) - \eta(t_1) \in V(t_1, r)^\perp$ imposed in Theorem 4.2.2.

<u>Theorem 4.3.1</u> Consider an extremal triple $(y(t), w(t), \mu(t))$. Let $y(t)$ be a neighboring extremal. Suppose that the relation $y(t) = x(t)$ holds on $[t_1, r]$. Then $w(t) = u(t)$ on $[t_1, r]$.

<u>Corollary 4.3.2</u> Suppose that $dim V(t_1, t)$ is a constant on $[t_1, t_2]$. Consider an extremal triple $(y(t), w(t), \mu(t))$ such that $y(t)$ is a neighboring extremal. If $y(t)$ coincides with $x(t)$ on an initial subinterval $[t_1, r]$ of the interval $[t_1, t_2]$ then $u(t) = w(t)$ on $[t_1, t_2]$.

In the case of the linear–quadratic control problem (1.4.1), (2.4.1) there exists a unique extremal trajectory which steers the initial point x_1 to the terminal point x_2 over the time interval $[t_1, t_2]$. The next theorem shows that the selected trajectory $x(t)$ has this property "in the small".

<u>Theorem 4.3.2</u> Consider an extremal triple $(y(t), w(t), \mu(t))$. If

    1. $y(t)$ is a neighboring extremal,

    2. $y(t_1) = x(t_1)$, $y(r) = x(r)$ for some $r \in [t_1, t_2]$,

then $y(t) = x(t)$ on $[t_1, r]$.

Proof. The first condition implies that $\mu(t_1) \in \Xi(r)$. On the other hand the relation $\Xi(r) \subset \Xi(t)$ holds for each t such that $t_1 \leq t < r$, hence, $\mu(t_1) \in \Xi(r)$. This completes the proof.

As it was shown by Example 1.10.1 the condition $|(x, \eta) - (y, \mu)|_{C[t_1, t_2]} < \epsilon^*$ cannot be removed. Hence, Theorem 4.3.2 has a local nature.

In the rest of this section we intend to derive our main statement (namely, the necessary and sufficient condition for branching) by using the above results. Consider $r \in [t_1, t_2]$ such that $dim V(t_1, r) < dim V(t_1, r + h)$ for some $h > 0$. Choose $\mu(t_1)$ as follows:

    1. the generated extremal trajectory $y(t)$ is a neighboring trajectory,

    2. $\mu(t_1) - \eta(t_1) \in V(t_1, r)^\perp$,

    3. $\mu(t_1) - \eta(t_1) \notin V(t_1, r + h)^\perp$.

Denote the extremal triple generated by $\mu(t_1)$ by $(y(t), w(t), \mu(t))$. Namely, $(y(t), \mu(t))$ is a solution of the Hamiltonian system (1.5.1), (1.5.2) with the initial conditions $(x(t_1), \mu(t_1))$. Due to Lemma 4.2.2 the extremal pairs $(x(t), u(t))$ and $(y(t), w(t))$ coincide on $[t_1, r]$. Suppose for a moment that $y(r+h) = x(r+h)$ for some positive h. This means that $\mu(t_1) \in \Xi(r+h)$ and $\mu(t_1) - \eta(t_1) \in V(t_1, r+h)^\perp$ – a contradiction to the third condition above.

The next statement follows from the previous discussion and Definition 1.4.2.

<u>Lemma 4.3.2</u> If $dim\, V(t_1, r) < dim\, V(t_1, r^+)$ then r is a branching point of the extremal trajectory $x(t)$. Namely, there exists an extremal trajectory $y(t)$ such that

$$y(t) = x(t) \text{ on } [t_1, r] \text{ and } x(t) \neq y(t) \text{ on } (r, t_2].$$

(Moreover, each ξ close to $\eta(t_1)$ and satisfying the relation

$$\xi - \eta(t_1) \in V(t_1, r)^\perp - V(t_1, r^+)^\perp$$

generates an extremal trajectory $y(t; \xi)$ which branches out of $x(t)$ at r.)

<u>Definition 4.3.2</u> We denote the set of discontinuity points of the function $dim V(t_1, s)$ on the interval $[t_1, t_2]$ by $\{r_i\}$.

We are now in a position to state and prove the main result of this section.

<u>Theorem 4.3.3</u> Each $r \in \{r_i\}$ is a branching point of the extremal trajectory $x(t)$. Let $y(t)$ be a neighboring extremal trajectory. If $y(t_1) = x(t_1)$ and $y(s) \neq x(s)$ for some $s \in [t_1, t_2]$, then there exists $r \in [t_1, s]$ such that the following conditions hold:

  1. $r \in \{r_i\}$,
  2. $y(t) = x(t)$ on $[t_1, r]$,
  3. $y(t) \neq x(t)$ on $(r, t_2]$.

<u>Corollary 4.3.3</u> The set of branching points of the extremal trajectory $x(t)$ is a finite set. The number of its elements is no more than the dimension of the state space R^n.

§4.4 Left branching

In this section we consider a set of left branching points of the selected extremal trajectory $x(t)$. Namely, we consider the points $l \in [t_1, t_2]$ such that there exists a neighboring extremal trajectory $y(t)$ which coincides with $x(t)$ on $[l, t_2]$, and pointwise different from it on $[t_2, l)$. We call such a point l a *left branching point* of $x(t)$, and say that $y(t)$ *branches out of* $x(t)$ at l.

The structure of the branching in this case is absolutely symmetric to the structure of branching considered in the previous section. To the ease of references in the last section we wish to state only the main theorems without proofs.

We first state the condition which ensures the absence of left branching points.

Theorem 4.4.1 Consider an extremal triple $(y(t), w(t), \mu(t))$ such that $|\mu(t_2) - \eta(t_2)|$ is small. Suppose that $V(t_2, l) = V(t_2, l - h)$ for some $h > 0$, and $w(t) = u(t)$ over $[l, t_2]$. Then $w(t) = u(t)$ over $[l - h, t_2]$.

We present below the necessary and sufficient conditions for left branching of the extremal trajectory $x(t)$.

Theorem 4.4.2 Consider an extremal triple $(y(t), w(t), \mu(t))$ such that $|\mu(t_2) - \eta(t_2)|$ is small. Suppose that $y(t) = x(t)$ on a certain subinterval $[l, t_2]$ of $[t_1, t_2]$. Then $w(t) = u(t)$ on $[l, t_2]$.

Lemma 4.4.1 If $dim\, V(t_2, l) < dim\, V(t_2, l^-)$, then l is a left branching point of the extremal trajectory $x(t)$. Namely, there is an extremal trajectory $y(t)$ such that

$$y(t) = x(t) \text{ on } [l, t_2] \text{ and } x(t) \neq y(t) \text{ on } [t_1, l).$$

(Moreover, each ξ close to $\eta(t_2)$ such that

$$\xi - \eta(t_2) \in V(t_2, l)^\perp - V(t_2, l^-)^\perp$$

generates an extremal trajectory $y(t; \xi)$ which branches out of $x(t)$ at l.)

Definition 4.4.1 We denote the set of discontinuity points of the function $dim V(t_2, s)$ on the interval $[t_1, t_2]$ by $\{l_j\}$.

Theorem 4.4.3 Each $l \in \{l_j\}$ is a left branching point of the extremal trajectory $x(t)$. Let $y(t)$ be a neighboring extremal trajectory. If $y(t_2) = x(t_2)$ and $y(s) = x(s)$ for some $s \in [t_1, t_2]$, then there exists $l \in \{l_j\}$ such that the following conditions holds:

    1. $l \in \{l_j\}$,

    2. $y(t) = x(t)$ on $[l, t_2]$,

    3. $y(t) \neq x(t)$ on $[t_1, l)$.

Corollary 4.4.1 The set of left branching points of the extremal trajectory $x(t)$ is finite. The number of its elements is no more than the dimension of the state space R^n.

§4.5 Two sided branching

In this section we investigate the set of branching pairs of the extremal trajectory $x(t)$. Namely, we consider pairs $\{l, r\}$ such, that there exists a neighboring extremal trajectory $y(t)$ which coincides with $x(t)$ on $[l, r]$, and pointwise different from it on the rest of the time interval. We intend to show that this set is some kind of mixture of two types of branching considered in the preceding sections. More precisely: a pair $\{l, r\}$ is a two sided branching pair of an extremal trajectory $x(t)$ if and only if the following condition holds:

$$dimV(l, r) < dimV(l, r^+) \quad \text{and} \quad dimV(r, l) < dimV(r, l^-). \tag{4.5.1}$$

Theorem 4.5.1 Suppose that condition (4.5.1) holds. Let $(y(t), w(t), \mu(t))$ be an extremal triple such that:

1. $x(r) = y(r)$,
2. $|\mu(t) - \eta(t)|$ is small on $[l, r]$,
3. $\mu(l) - \eta(l) \in V(l, r)^{\perp} - V(l, r^+)^{\perp}$,
4. $\mu(r) - \eta(r) \in V(r, l)^{\perp} - V(r, l^-)^{\perp}$.

Then the extremal trajectory $y(t)$ branches out of $x(t)$ at $\{l, r\}$. Namely, $y(t) = x(t)$ on $[l, r]$ and $y(t) \neq x(t)$ for $t \in [t_1, l) \cup (r, t_2]$.

Proof. First, by using Lemma 4.4.1 we obtain that

$$x(t) = y(t) \text{ on } [l, r] \text{ and } x(t) \neq y(t) \text{ on } [t_1, l).$$

Next, by Lemma 4.3.2 $x(t) \neq y(t)$ on $(r, t_2]$. This completes the proof.

Theorem 4.5.2 If there exists an extremal triple $(y(t), w(t), \mu(t))$ such that

1. $y(t)$ is a neighboring extremal trajectory,
2. $x(t) = y(t)$ on $[l, r]$,
3. $y(t)$ is pointwise different from $x(t)$ on the rest of the time interval $[t_1, t_2]$,

then the condition (4.5.1) holds and

1. $\mu(l) - \eta(l) \in V(l, r)^{\perp} - V(l, r^+)^{\perp}$,
2. $\mu(r) - \eta(r) \in V(r, l)^{\perp} - V(r, l^-)^{\perp}$.

Proof. Theorem 4.3.2 implies that $dimV(l, r) < dimV(l, r^+)$. Theorem 4.4.3 yields that $dimV(r, l) < dimV(r, l^-)$. We now employ Corollary 3.3.2 and this completes the proof.

Corollary 4.5.1 A pair $\{l, r\}$ is a branching pair of the extremal trajectory $x(t)$ if and only if the condition (4.5.1) holds.

In accordance with Lemma 3.3.2 we obtain the following:

Corollary 4.5.2 The set of two sided branching pairs is at most countable.

Remark 4.5.1 For the sake of technical convenience, one can change Hypothesis 1.3.1 by the following stronger Hypothesis 4.5.1:

Hypothesis 4.5.1 There exists a negative definite $(n+m) \times (n+m)$ matrix Q such that the following relation holds along $x(t)$

$$\begin{pmatrix} H_{xx}(t, x(t), u(t), \eta(t)) & H_{ux}(t, x(t), u(t), \eta(t)) \\ H_{xu}(t, x(t), u(t), \eta(t)) & H_{uu}(t, x(t), u(t), \eta(t)) \end{pmatrix} \leq Q,$$

and the matrix

$$\begin{pmatrix} H_{xx}(t, x(t), u(t), \eta(t)) & H_{ux}(t, x(t), u(t), \eta(t)) \\ H_{xu}(t, x(t), u(t), \eta(t)) & H_{uu}(t, x(t), u(t), \eta(t)) \end{pmatrix}$$

is uniformly continuous with respect to t (see Hypothesis 1.3.1).

We wish to emphasize that the verification of Hypothesis 4.5.1 is much easily than that of Hypothesis 1.3.1.

Chapter 5

Linear systems with vector valued performance index

§5.1 Preface

This chapter concerns the linear control system

$$\dot{x}(t) = A(t)x(t) + B(t)u(t), \tag{5.1.1}$$

with the following k-dimensional vector cost:

$$c(x, u) = (\, c_1(x, u), \ldots\ldots, c_k(x, u)\,), \tag{5.1.2}$$

where for each $i = 1, \ldots, k$ $c_i(x, u) = \int_{t_1}^{t_2} f_i(t, x(t), u(t))dt$ and $f_i(t, x, u)$ is assumed here to be a strictly convex function in (x, u).

The notion of a Pareto pair and an extended extremal trajectory have been introduced in section 1.6, Chapter 1. The connection between the notions also has been discussed there. In this chapter we intend to investigate the structure of the set of branching points in optimal control problem (5.1.1), (5.1.2). To this end the auxiliary definitions are introduced as follows:

Let $P^+ = \{\, c \in R^k \mid c_i \geq 0 \text{ all } i, \; c_j > 0 \text{ some } j\,\}$ and $P^o = \{\, c \in R^k \mid c_i > 0 \text{ all } i\,\}$.

<u>Definition 5.1.1</u> The linear control system (5.1.1) with the following μ–scalar cost:

$$c_\mu(x, u) = \sum_{i=1}^{k} \mu_i c_i(x, u), \quad \mu \in P^+ \tag{5.1.3}$$

is called a μ-control problem.

By a slight modification of Theorem 8 (see [11], p. 209) one can show that the extended attainable set of the control problem (5.1.1) − (5.1.2) (see Definition 1.6.7) is a closed, convex set. Then, by using a modification of Theorem 8.3.1 the following statement can be proved:

<u>Theorem 5.1.1</u> An admissible pair $(x(t), u(t))$ is a Pareto pair if and only if $(x(t), u(t))$ is an extremal pair of a μ-control problem for some $\mu \in P^+$.

Our main goal in this chapter is a description of branching points of an extended extremal trajectory $(c_u(t), x(t))$ (see Definition 1.6.4). Namely, we consider the points r such that there exists an extended extremal trajectory $(c_w(t), y(t))$ which satisfies the following condition:

$$(c_u(t), x(t)) = (c_w(t), y(t)) \text{ on } [t_1, r] \text{ and } (c_u(t), x(t)) \neq (c_w(t), y(t)) \text{ on } (r, t_2].$$

If $x(t)$ and $y(t)$ are the extremal trajectories of the same μ–control problem and $(c_w(t), y(t))$ branches out of $(c_u(t), x(t))$ at r, then due to Theorem 2.3.4, $r \in R\{[t_1, t_2]\}$. In the next section we investigate the branching points of $(c_u(t), x(t))$ generated by extended extremals $(c_w(t), y(t))$, such that $y(t)$ is an extremal trajectory of λ–control problem and, $\lambda \neq \mu$.

§5.2 Vector branching

Our main goal in this section is a description of branching points under the condition

$$dimV(t_1, s) \text{ is a constant on } [t_1, t_2].$$

Existence of such a branching is the first question we wish to discuss. To this end we introduce the following example.

Example 5.2.1

Consider the linear control system

$$\dot{x}(t) = u(t), \quad x(0) = 0,$$

with a vector cost functional

$$c_1(x, u) = \int_0^2 u^2(t)dt, \quad c_2(x, u) = \int_0^2 u^2(t) + \varphi(t)x^2(t)dt.$$

Where x, u–are scalars, $\varphi(t)$ is a real valued function which is defined as follows:

$$\varphi(t) = 0 \text{ on } [0, 1], \quad \varphi(t) = 1 \text{ on } (1, 2].$$

It is easy to check that the pairs $(x(t), u(t)) = (t, 1)$ and $(y(t), w(t))$ which is defined as follows:

$$(y(t), w(t)) = (t, 1) \text{ on } [0, 1], \quad (y(t), w(t)) = (e^{t-1}, e^{t-1}) \text{ on } (1, 2],$$

are Pareto pairs which coincide on $[0, 1]$ and are pointwise different on $(1, 2]$. Note that in this case $dimV(0, t) = 1$ for each $t \neq 0$.

As it was mentioned in the preceding section such a branching can not take place for Pareto trajectories $x(t)$, $y(t)$ which are extremal trajectories of the same μ–control problem. Indeed, in the above example, the first trajectory is an extremal trajectory of $(1, 0)$–control problem, and the second one is an extremal trajectory of $(0, 1)$–control problem.

In this chapter we consider a certain Pareto pair $(x(t), u(t))$. As it was mentioned in Theorem 5.1.1, there is $\mu \in P^+$ such that $(x(t), u(t))$ is an extremal pair of μ–control problem.

Tangent space to the cost set.

In this subsection the equation of the tangent space to the attainable cost set will be derived. This tangent space together with the attainable set, determines the set of branching point of the extended extremal trajectory $(c_u(t), x(t))$.

Definition 5.2.1 Let $O(t_1, s)$ be a set of all admissible controls $v(t)$ which steers 0 to 0 during the time $[t_1, s]$, namely

$$O(t_1, s) = \{ \, v \mid v \in U[t_1, s], \, \int_{t_1}^{s} \Phi(s, t) B(t) v(t) dt = 0. \, \}$$

Here $\Phi(t, t_1)$ is the transition matrix of $\frac{d}{dt} x = A(t)x$, i.e. $x(t) = \Phi(t, t_1) x_1$ is the solution of $\frac{d}{dt} x = A(t)x$ with $x(t_1) = x_1$.

Remark 5.2.1 An admissible control $w(t)$ steers $x(t_1)$ to $x(t_2)$ if and only if $w \in u + O(t_1, t_2)$.

Define the linear transformations $T_i : O(t_1, t_2) \mapsto R^1$, $i = 1, ..., k$ as follows:

$$T_i(v) = \int_{t_1}^{t_2} \{ \, \frac{\partial f_i}{\partial x}(t, x(t), u(t)) (\int_{t_1}^{t} \Phi(t, s) B(s) v(s) ds) + \frac{\partial f_i}{\partial u}(t, x(t), u(t)) v(t) \, \} dt.$$

Note, that ($T_1(w - u), ..., T_k(w - u)$) is a linear estimation of the difference $c(y, w) - c(x, u)$ for $w \in u + O(t_1, t_2)$. Denote the linear mapping $T_1, .., T_k : O(t_1, t_2) \to R^k$ by T, namely $T(v) = (T_1(v), ..., T_k(v))$.

Definition 5.2.2 Denote the $Im\ T$ by $W(t_1, t_2; u)$ or, if it does not lead to ambiguity, simply by $W(t_1, t_2)$.

Lemma 5.2.1 If $(x(t), u(t))$ is a Pareto pair then:

1. $W(t_1, t_2) \cap P^o = \emptyset$,
2. there exists $\mu \in P^+$ such that $\mu \in W(t_1, t_2)^\perp$,
3. for each $\mu \in P^+ \bigcap W(t_1, t_2)^\perp$ and $w \in u + O(t_1, t_2)$ the following relation holds

$$\langle \mu, c(y, w) - c(x, u) \rangle \; > 0.$$

Our proof consists of three steps. First we prove that $W(t_1, t_2) \cap P^o = \emptyset$, next we fin$\cdot$ $\mu \in P^+ \cap W(t_1, t_2)^\perp$, and in the last step shows that $\langle \mu, c(y, w) - c(x, u) \rangle \; > 0$.

Proof. Step 1. Due to the condition that $(x(t), u(t))$ is a Pareto pair we obtain the followin$\cdot$ relation:

for each $w \in u + O(t_1, t_2)$ $\quad c_i(x, u) \leq c_i(y, w) \quad i = 1, .., k.$

Suppose that there exists $c \in W(t_1, t_2) \cap P^o$. This implies that $c_i > 0$ each i. Let $v \in O(t_1, t_2)$ such that $T(v) = -c$. Consider a family of controls $w_\alpha(t)$ such that $w_\alpha(t) = u(t) + \alpha v(t)$, α is a real number. Note that

$$w_\alpha \in u + O(t_1, t_2) \quad \text{all} \quad \alpha,$$

$$c_i(y_\alpha, w_\alpha) - c_i(x, u) = \alpha[T(v)]_i + o|\alpha| = -\alpha c_i + o|\alpha| \quad \text{all} \quad i.$$

Hence, for α small enough, $c_i(y_\alpha, w_\alpha) - c_i(x, u) < 0$ all i – a contradiction.

Step 2. There exists a linear functional μ which separates two disjoint convex sets $W(t_1, t_2)$ and P^o, namely

$$\langle w, \mu \rangle \leq \langle p, \mu \rangle \text{ for each } w \in W(t_1, t_2) \text{ and } p \in P^o.$$

(See e.g., Rudin [18], p. 58, Theorem 3.4.) It is clear that μ must be an element of $W(t_1, t_2)^\perp$, and may be chosen in a way such that $\mu \in P^+$.

Step 3. By using the following well known property of strictly convex functions:

$$f(t, y, w) - f(t, x, u) > \frac{\partial f}{\partial x}(t, x, u)(y - x) + \frac{\partial f}{\partial u}(t, x, u)(w - u),$$

we obtain

$$\langle \mu, c(y, w) - c(x, u) \rangle > \langle \mu, T(w - u) \rangle = 0.$$

This completes the proof.

We show now the connection between the tangent spaces to the cost sets of the extended extremal trajectories which coincide on an initial subinterval $[t_1, s]$. Consider another Pareto extremal pair $(y(t), w(t))$ and suppose that:

1. $(y(t), w(t))$ is an extremal pair of λ–control problem on $[t_1, t_2]$, $\mu \neq \lambda$, $\lambda \in P^+$,
2. $(y(t), w(t)) = (x(t), u(t))$ for $t \in [t_1, s]$.

The second condition immediately yields that $W(t_1, s; u) = W(t_1, s; w)$. Theorem 5.1.1 implies that $\lambda, \mu \in W(t_1, s)^\perp$. Consider $dimW(t_1, s; u)$ as a function of s. Choose s_1, s_2 such that $t_1 \leq s_1 \leq s_2$ holds. One can identifies $O(t_1, s_1)$ with

$$O(t_1, s_2) \bigcap \{ v \in O(t_1, s_2) \mid v(t) = 0 \text{ for } t \in (s_1, s_2] \}.$$

Definition 5.2.2 implies that $W(t_1, s_1) \subset W(t_1, s_2)$. On the other hand for each s the space $W(t_1, s)$ is a subspace of R^k and, therefore, $dimW(t_1, s) \leq k$. This consideration shows that $dimW(t_1, s)$ is a monotone non decreasing step function with respect to s . This implies that the number of its discontinuity points is finite. We introduce now the following definition:

Definition 5.2.3 For each scalar $s \in [t_1, t_2]$ denote by $W(t_1, s)$ the following vector space

$$W(t_1, s^+) = \bigcap_{h > 0} W(t_1, s + h).$$

In the rest of the section we investigate the branching points of extended Pareto trajectories. On the other hand, we will show, by examples, the distinction between branching of the Pareto trajectories and that of the extended Pareto trajectories.

Necessary and sufficient conditions for branching.

We begin now to derive the necessary and sufficient conditions for branching of the extended extremal trajectory $(c_u(t), x(t))$.

<u>Theorem 5.2.1</u> Consider the extended Pareto trajectory $(c_u(t), x(t))$. Assume that for some positive scalar h and $s \in [t_1, t_2]$

$$dimV(t_1, s) = dimV(t_1, s + h), \quad dimW(t_1, s) = dimW(t_1, s + h)$$

and there is an extended Pareto trajectory $(c_w(t), y(t))$ which coincides with $(c_u(t), x(t))$ on $[t_1, s]$. Then $(c_w(t), y(t)) = (c_u(t), x(t))$ on $[t_1, s + h]$.

<u>Proof.</u> There exists $\mu \in P^+ \bigcap W(t_1, s; w)^\perp$ such that $y(t)$ is an extremal trajectory of μ-control problem on $[t_1, t_2]$ (see Lemma 5.2.1). The relation

$$W(t_1, t_2; w)^\perp \subset W(t_1, s; w)^\perp$$

implies that $\mu \in W(t_1, s; w)^\perp$. The relation $(c_u(t), x(t)) = (c_w(t), y(t))$ implies that $W(t_1, s; u) = W(t_1, s; w)$, hence $\mu \in W(t_1, s; u)^\perp$. On the other hand, due to the assumption, $W(t_1, s; u) = W(t_1, s + h; u)$ and, therefore, $\mu \in W(t_1, s + h; u)^\perp$. Applying Lemma 5.2.1 again we obtain that $x(t)$ is an extremal trajectory of μ-control problem on $[t_1, s + h]$. The result follows now from Theorem 2.3.4.

We wish now to emphasize the difference between the notions of an extremal trajectory and an extended extremal trajectory. To this end the following example is introduced.

<u>Example 5.2.2</u> Branching of extremal trajectories under the condition

$$dimV(0, t) = dimW(0, t; u) = constant.$$

Consider the linear control system

$$\dot{x}(t) = u_1(t) + u_2(t), \quad x(0) = 0,$$

with a vector cost functional

$$c_1(x, u) = \int_0^2 u_1^2(t) + u_2^2(t)dt, \quad c_2(x, u) = \int_0^2 \varphi(t)u_1^2(t) + u_2^2(t)dt.$$

Where x is a scalar, $u \in R^2$, and $\varphi(t)$ is a real valued function defined as follows:

$$\varphi(t) = 2 \text{ on } [0, 1], \quad \varphi(t) = 3 \text{ on } (1, 2].$$

Consider an extremal pair $(x(t), u(t))$ of $(1, 0)$–control problem which defined as follows:

$$x(t) = 3t, \quad u(t) = (\frac{3}{2}, \frac{3}{2}).$$

In this case

$$dimV(0, t) = 1, \quad dimW(0, t; u) = 1 \quad \text{for each } t \in (0, 2].$$

On the other hand a pair $(y(t), w(t))$ which is defined as follows:

$$y(t) = 3t, \quad u(t) = (1, 2) \quad \text{on} \quad [0, 1], \quad y(t) = 3 + \frac{8}{3}(t - 1), \quad u(t) = (\frac{2}{3}, 2) \quad \text{on} \quad (1, 2]$$

is an extremal pair of $(0, 1)$–control problem. The corresponding trajectories $y(t)$, $x(t)$ coincide on $[0, 1]$, and are pointwise different on $(1, 2]$.

<u>Theorem 5.2.2</u> Consider an extended Pareto trajectory $(c_u(t), x(t))$. Let us suppose that for some $s \in [t_1, t_2]$

$$\text{either } dimV(t_1, s) < dimV(t_1, s^+) \quad \text{or} \quad dimW(t_1, s) < dimW(t_1, s^+).$$

Then there is an extended Pareto trajectory $(c_w(t), y(t))$ which branches out of $(c_u(t), x(t))$ at s.
<u>Proof.</u> If $dimV(t_1, s) < dimV(t_1, s^+)$, then the desired result follows immediately from Theorem 2.3.3. We assume, therefore, that

$$dimV(t_1, s) = dimV(t_1, s^+) \quad \text{and} \quad dimW(t_1, s; u) < dimW(t_1, s^+; u).$$

Choose $\mu \in P^+ \bigcap \{W(t_1, s; u)^\perp - W(t_1, s^+; u)^\perp\}$. Choose an admissible pair $(y(t), w(t))$ such that $x(t_1) = y(t_1)$, $x(s) = y(s)$, and $y(t)$ is an extremal trajectory of μ–control problem. Due to Lemma 5.2.1, $(x(t), u(t))$ is an extremal pair of μ–control problem on $[t_1, s]$, this implies that $(x(t), u(t)) = (y(t), w(t))$ on $[t_1, s]$. For each $\epsilon > 0$ there exists a positive h, such that $\epsilon > h > 0$ and

$$(c_u(s + h), x(s + h)) \neq (c_w(s + h), y(s + h))$$

(otherwise $\mu \in P^+ \bigcap \{ W(t_1, s + \epsilon; u)^\perp \}$ -a contradiction), this implies that

$$(c_u(t), x(t)) \neq (c_w(t), y(t)) \quad \text{on} \quad [s + h, t_2].$$

The arbitrariness of the ϵ yields

$$(c_u(t), x(t)) \neq (c_w(t), y(t)) \quad \text{on} \quad (s, t_2],$$

his completes the proof.

<u>Corollary 5.2.1</u> Consider an extended Pareto trajectory $(c_u(t), x(t))$. We suppose that for some $s \in [t_1, t_2]$

$$dimW(t_1, s) < dimW(t_1, s^+).$$

Then for each

$$\mu \in P^+ \bigcap \{ \ W(t_1, s; \boldsymbol{u})^\perp - W(t_1, s^+; \boldsymbol{u})^\perp \ \}$$

there exists an extended Pareto trajectory $(c_w(t), y(t))$ such that:

1. $y(t)$ is an extremal trajectory of μ–control problem,
2. $(c_w(t), y(t)) = (c_u(t), x(t))$ on $[t_1, s]$, and $(c_w(t), y(t)) \neq (c_u(t), x(t))$ on $(s, t_2]$.

We wish to emphasize that Corollary 5.2.1 fails for Pareto trajectories (see Example 1.10.3).

The next result follows from Theorem 5.2.1 and Theorem 5.2.2.

<u>Corollary 5.2.2</u> The set of branching points of an extended Pareto trajectory is a finite set. The # of its elements is no more than $n + k - 1$. (Here n is the dimension of the state space, k is the dimension of the cost space.)

We wish to mention that the set of branching points of an extended Pareto trajectory, in contrast with that of an extremal trajectory, depends on the matrices $A(t)$, $B(t)$, the cost functional, namely the functions $f_1(t, x, u), .., f_k(t, x, u)$ and the extended Pareto trajectory (see Example 1.10.4).

§5.3 Branching pairs

Left branching points.

In this subsection we investigate left branching points of an extended Pareto trajectory $(c_u(t), x(t))$. Namely, we consider points $l \in [t_1, t_2]$ such, that there exists an extended Pareto trajectory $(c_w(t), y(t))$, and the following relation holds:

$$(c_u(t), x(t)) = (c_w(t), y(t)) \text{ on } [l, t_2] \text{ and } (c_u(t), x(t)) \neq (c_w(t), y(t)) \text{ on } [t_1, l).$$

The structure of the left branching is absolutely symmetric to the structure of the branching that has been considered in the preceding section. For this reason we wish to display the necessary definitions and to state the main theorems without proofs.

Definition 5.3.1 For each scalar $s \in [t_1, t_2]$ we define the vector space $W(t_2, s^-)$ by the following formula:

$$W(t_2, s^-) = \bigcap_{h > 0} W(t_2, s - h)$$

Theorem 5.3.1 Consider the extended Pareto trajectory $(c_u(t), x(t))$. Suppose that for some positive scalar h and $s \in [t_1, t_2]$ there exists an extended Pareto trajectory $(c_w(t), y(t))$ which coincides with $(c_u(t), x(t))$ on $[s, t_2]$, and the following condition holds:

$$dimV(t_2, s) = dimV(t_2, s - h) \quad \text{and} \quad dimW(t_2, s) = dimW(t_2, s - h).$$

Then $(c_w(t), y(t)) = (c_u(t), x(t))$ on $[s - h, t_2]$.

Theorem 5.3.2 Consider the extended Pareto trajectory $(c_u(t), x(t))$. If for some $s \in [t_1, t_2]$

$$\text{either} \quad dimV(t_2, s) < dimV(t_2, s^-) \quad \text{or} \quad dimW(t_2, s) < dimW(t_2, s^-),$$

then there exists an extended Pareto trajectory $(c_w(t), y(t))$ which branches out $(c_u(t), x(t))$ at s.

Corollary 5.3.1 Consider the extended Pareto trajectory $(c_u(t), x(t))$. If for some $s \in [t_1, t_2]$

$$dimW(t_2, s) < dimW(t_2, s^-),$$

then for each

$$\mu \in P^+ \bigcap \{ \, W(t_2, s; u)^\perp - W(t_2, s^-; u)^\perp \, \}$$

there is an extended Pareto trajectory $c_w(t), y(t)$ such that

1. $y(t)$ is an extremal trajectory of μ–control problem,
2. $(c_w(t), y(t)) = (c_u(t), x(t))$ on $[s, t_2]$, $(c_w(t), y(t)) \neq (c_u(t), x(t))$ on (t_1, s).

Corollary 5.3.2 The set of left branching points of an extended Pareto trajectory is a finite set. The # of its elements is no more than $n + k - 1$. (Here n is a dimension of a state space, k is a dimension of a cost space.)

The two sided branching.

In this subsection we investigate the set of branching pairs of the extended Pareto trajectory $c_u(t), x(t))$ which is defined on the time interval $[t_1, t_2]$. Namely, we consider the set of pairs $\{l, r\}$ such that, there exists an extended Pareto trajectory $(c_w(t), y(t))$ which satisfies the following:

$$(c_u(t), x(t)) = (c_w(t), y(t)) \text{ on } [l, r] \quad \text{and} \quad (c_u(t), x(t)) \neq (c_w(t), y(t)) \text{ on } [t_1, l) \cup (r, t_2].$$

As one can expect, this set is some kind of mixture of the "scalar" branching and the "vector" branching (see section 5.2, example 5.2.1). More precisely:

Consider an extended Pareto trajectory $(c_u(t), x(t))$ and the following list of conditions:

1. $dimV(r,l) < dimV(r,l^-)$ and $dimV(l,r) < dimV(l,r^+)$,
2. $dimV(r,l) < dimV(r,l^-)$ and $dimW(l,r) < dimW(l,r^+)$,
3. $dimW(r,l) < dimW(r,l^-)$ and $dimV(l,r) < dimV(l,r^+)$,
4. $dimW(r,l) < dimW(r,l^-)$ and $dimW(l,r) < dimW(l,r^+)$.

<u>Theorem 5.3.3</u> If at least one of the listed above conditions $1 - 4$ holds than $\{l,r\}$ is a branching pair of the extended Pareto trajectory $(c_u(t), x(t))$.

<u>Proof.</u> **Condition 1.** In this case the result follows from Theorem 3.3.2.

Condition 2. Choose

$$\mu \in P^+ \bigcap \{ \ W(l,r;u)^\perp - W(l,r^+;u)^\perp \ \},$$

the trajectory $x(t)$ is an extremal trajectory of μ–control problem on $[l,r]$. Hence (due to Theorem 5.2.2), there exists an extended Pareto trajectory $(c_{w_1}(t), y_1(t))$ on $[l,t_2]$ such that

$$(c_{w_1}(t), y_1(t)) = (c_u(t), x(t)) \ \text{ on } \ [l,r], \ \ (c_{w_1}(t), y_1(t)) \neq (c_u(t), x(t)) \ \text{ on } \ (r, t_2].$$

Now we wish to extend $(c_{w_1}(t), y_1(t))$ on the whole interval $[t_1, t_2]$. Choose an extremal trajectory $y_2(t)$ of μ–control problem in such a way that $y_2(l) = x(l)$, $y_2(r) = x(r)$. If $y_2(t) \neq x(t)$ on $[t_1, l)$ then define $y(t)$ as follows:

$$y_1(t) = y(t) \ \text{ on } \ [l,t_2], \ \ y_2(t) = y(t) \ \text{ on } \ [t_1, l).$$

If $y_2(t) = x(t)$ on $[t_1, l)$ then, due to Lemma 3.3.2, there exists an extremal trajectory $y_3(t)$ of μ–control problem such that

$$y_3(t) \neq y_2(t) = x(t) \ \text{ on } \ [t_1, l).$$

In this case we define $y(t)$ as follows:

$$y_1(t) = y(t) \ \text{ on } \ [l,t_2], \ \ y_3(t) = y(t) \ \text{ on } \ [t_1, l).$$

The generated trajectory $y(t)$ is an extremal trajectory of μ–control problem. This implies that $(c_w(t), y(t))$ is the desired extended Pareto trajectory.

Condition 3. Absolutely symmetric to Condition 2.

Condition 4. Choose

$$\mu \in P^+ \bigcap \{ \ W(l,r;u)^\perp - W(l,r^+;u)^\perp \ \}$$

Choose an extremal trajectory $y(t)$ of μ–control problem such that

$$y(l) = x(l), \quad y(r) = x(r).$$

Due to Corollary 5.2.1 $y(t) \neq x(t)$ on $(r, t_2]$, due to Corollary 5.3.1 $y(t) \neq x(t)$ on $[t_1, l)$. The trajectories $x(t)$, $y(t)$ are extremal trajectories of the μ–control problem on $[l, r]$, this implies that

$$x(t) = y(t) \quad \text{on} \quad [l, r].$$

This completes the proof.

<u>Theorem 5.3.4</u> If $\{l, r\}$ is a branching pair of the extended Pareto trajectory $(c_u(t), x(t))$ then at least one of the conditions 1–4 holds.
<u>Proof.</u> The proof follows from Lemma 3.3.2, Theorem 5.2.1 and Theorem 5.3.1.

The two last theorems yield the main result of this section, namely the necessary and sufficient conditions for branching of the extended Pareto trajectory.
<u>Theorem 5.3.5</u> One of the listed above conditions $1 - 4$ holds if and only if $\{l, r\}$ is a branching pair of the extended Pareto trajectory $(c_u(t), x(t))$.

Chapter 6

Nonlinear control systems with vector cost

§6.1 Introduction

This chapter is devoted to the investigation of branching points of extended extremal trajectories in the nonlinear control system

$$\dot{x}(t) = F(t, x(t), u(t)) \tag{6.1.1}$$

with k-dimensional vector cost

$$c(x, u) = (\ c_1(x, u), ..., c_k(x, u)\ ), \text{ each } c_i(x, u) = \int_{t_1}^{t_2} f_i(t, x(t), u(t)) dt. \tag{6.1.2}$$

Consider an extended extremal trajectory $(c_u(t), x(t))$ (see Definition 1.6.4). We say that an extended extremal trajectory $(c_w(t), y(t))$ is a neighboring extended extremal trajectory if

$$\max_{t_1 \leq t \leq t_2} |(c_u(t), x(t), \eta(t), \mu) - (c_w(t), y(t), \xi(t), \lambda)| < \epsilon^{**}.$$

The value of the ϵ^{**} will be determined in Definition 6.3.1. A point r is a branching point of the extended extremal trajectory $(c_u(t), x(t))$ if there exists a neighboring extended extremal trajectory $(c_w(t), y(t))$ such that

$$(c_u(t), x(t)) = (c_w(t), y(t)) \text{ on } [t_1, r] \text{ and } (c_u(t), x(t)) \neq (c_w(t), y(t)) \text{ on } (r, t_2].$$

The underlying idea is the following: The last n coordinates $x(t)$ of the extended extremal trajectory $(c_u(t), x(t))$ is an extremal trajectory of the control system (6.1.1) with some μ–scalar cost

$$c_\mu(x, u) = \sum_{i=1}^{k} \mu_i c_i(x, u), \ \mu_i \geq 0 \text{ for all } i, \ \mu_j > 0 \text{ for some } j. \tag{6.1.3}$$

Namely, one can represent $x(t)$ as a solution of the Hamiltonian system

$$\dot{x}(t) = H_\eta(t, x(t), u(t), \eta(t), \mu) \tag{6.1.4}$$

$$\dot{\eta}(t) = -H_x(t, x(t), u(t), \eta(t), \mu) \tag{6.1.5}$$

where the Hamiltonian H is defined as follows:

$$H(t, x, u, \eta, \mu) = -\sum_{i=1}^{k} \mu_i f_i(t, x, u) + \eta F(t, x, u)$$

(see Definition 1.6.2).

<u>Definition 6.1.1</u> The system (6.1.4)–(6.1.5) is called the μ–Hamiltonian system.

By minor changes of the considerations which have been presented in Chapter 4, section 4.1 one can show the existence of a positive constant ϵ_1, such that each solution $(y(t), \xi(t))$ of a λ–Hamiltonian system with

$$|(y(t_1), \xi(t_1), \lambda) - (x(t_1), \eta(t_1), \mu)| < \epsilon_1$$

is an extremal pair of the control system (6.1.1) with λ–scalar cost. The idea of representation of extremal trajectories as solutions of the family of Hamiltonian systems allows to apply the ordinary branching machinery which has been developed in Chapter 4.

The chapter is organized as follows: In the second section the necessary and sufficient conditions for matching of extremal pairs are derived. In section 3 we establish the necessary and sufficient conditions for matching of extended extremal trajectories. This is the main part of the chapter. In section four the set of branching pairs of an extended extremal trajectory is described.

§6.2 Matching of Pareto pairs

We consider in this chapter an extremal quadruple $(x(t), u(t), \eta(t), \mu)$ of the optimal control problem (6.1.1), (6.1.2) (see Definition 1.6.1). Our main goal in this section is the description of the conditions which must be imposed on the initial values (ξ, λ) in order to fulfill the relation

$$(x(t; \xi, \lambda), u(t; \xi, \lambda)) = (x(t; \eta, \mu), u(t; \eta, \mu))$$

on an initial subinterval $[t_1, r]$ of the time interval $[t_1, t_2]$. Here $x(t; \xi, \lambda)$ is the first n coordinates of a solution $(x(t), \eta(t))$ of the λ–Hamiltonian system (6.1.4)–(6.1.5) with $x(t_1) = x_1$, $\eta(t_1) = \xi$ and $u(t; \xi, \lambda) = u(t; x(t), \eta(t), \lambda)$ (see Definition 1.6.1).

The necessary and sufficient conditions for the branching of the extended extremal trajectory $c_u(t), x(t))$ are derived now. Consider an extended extremal trajectory $(c_w(t), y(t))$ such that $y(t)$ the extremal trajectory of a λ–control problem. Namely, there is an absolutely continuous vector function $\xi(t)$ such that $(y(t), w(t), \xi(t))$ is an extremal triple of the λ–control problem. Suppose that

$$(y(t), w(t)) = (x(t), u(t)) \quad \text{on} \quad [t_1, r].$$

this yields

$$\frac{d}{dt}(\xi(t) - \eta(t)) = \frac{\partial}{\partial x} f_\gamma(t, x(t), u(t)) - (\xi(t) - \eta(t))\frac{\partial}{\partial x} F(t, x(t), u(t)).$$

Where $\gamma = \lambda - \mu$, $f(t, x, u) = (f_1(t, x, u), ..., f_k(t, x, u))$, and $f_\gamma(t, x, u) = \langle \gamma, f(t, x, u)\rangle$. We denote $\xi(t) - \eta(t)$ by $\Delta(t)$ and obtain the following result:

<u>Lemma 6.2.1</u> If the relation $(c_w(t), y(t), w(t)) = (c_u(t), x(t), u(t))$ holds on $[t_1, r]$ for extended extremal trajectory $(c_w(t), y(t))$ then

$$\Delta(t) = \Delta(t_1)\Phi(t_1, t) + \int_{t_1}^t \frac{\partial f_\gamma}{\partial x}(\sigma, x(\sigma), u(\sigma))\Phi(\sigma, t)d\sigma \quad \text{on} \quad [t_1, r].$$

where $\Phi(t, t_1)$ is the transition matrix of $\frac{d}{dt}\phi = A(t)\phi$, i.e. $\phi(t) = \Phi(t, t_1)x_1$ is the solution of the equation $\frac{d}{dt}\phi = A(t)\phi$ with $\phi(t_1) = x_1$.

Now we wish to determine conditions which must be imposed on a pair (Δ, γ) in order to fulfill the relation

$$u(t; \eta + \Delta, \mu + \gamma) = u(t; \eta, \mu) \quad \text{on} \quad [t_1, r]. \tag{6.2.1}$$

(Namely, the triple $(x(t), u(t), \eta(t) + \Delta(t))$ is an extremal triple of $(\mu + \gamma)$−control problem.)

Suppose, therefore, that the relation (6.2.1) indeed holds. This relation implies, due to Lemma 6.2.1, that

$$\Delta(t) = \Delta(t_1)\Phi(t_1, t) + \int_{t_1}^t \frac{\partial f_\gamma}{\partial x}(\sigma, x(\sigma), u(\sigma))\Phi(\sigma, t)d\sigma \quad \text{on} \quad [t_1, r].$$

On the other hand, the Pontryagin Maximum Principle yields the following:

$$\frac{\partial}{\partial u}\{ -f_{\mu+\gamma}(t, x(t), u(t)) + (\eta(t) + \Delta(t))F(t, x(t), u(t)) \} = 0 \text{ on } [t_1, r],$$

hence

$$\frac{\partial}{\partial u}\{ -f_\gamma(t, x(t), u(t)) + \Delta(t)F(t, x(t), u(t)) \} = 0 \text{ on } [t_1, r].$$

Namely, for each $w \in U[t_1, r]$ the following relation holds on $[t_1, r]$:

$$\{ \Delta(t_1)\Phi(t_1, t)B(t) + (\int_{t_1}^t \frac{\partial f_\gamma}{\partial x}(\sigma, x(\sigma), u(\sigma))\Phi(\sigma, t)d\sigma)B(t) - \frac{\partial f_\gamma}{\partial u}(t, x(t), u(t)) \}w(t) = 0.$$

This yields

$$\int_{t_1}^r \Delta(t_1)\Phi(t_1, t)B(t)w(t)dt +$$

$$+ \int_{t_1}^r \{ (\int_{t_1}^t \frac{\partial f_\gamma}{\partial x}(\sigma, x(\sigma), u(\sigma))\Phi(\sigma, t)d\sigma)B(t) - \frac{\partial f_\gamma}{\partial u}(t, x(t), u(t)) \}w(t)dt = 0. \tag{6.2.?}$$

Introduce now $n \times m$ and $k \times m$ matrices $\Lambda_1(t)$ and $\Lambda_2(t)$ as follows:

$$\Lambda_1(t) = \Phi(t_1, t)B(t)$$

and

$$\Lambda_2(t) = \begin{pmatrix} (\int_{t_1}^{t} \frac{\partial f_1}{\partial x}(\sigma, x(\sigma), u(\sigma))\Phi(\sigma, t)d\sigma)B(t) - \frac{\partial f_1}{\partial u}(t, x(t), u(t)) \\ \cdots\cdots \\ \cdots\cdots \\ (\int_{t_1}^{t} \frac{\partial f_k}{\partial x}(\sigma, x(\sigma), u(\sigma))\Phi(\sigma, t)d\sigma)B(t) - \frac{\partial f_k}{\partial u}(t, x(t), u(t)) \end{pmatrix}$$

Equation (6.2.2) implies that

$$\langle(\Delta, \gamma), (\int_{t_1}^{r} \Lambda_1(t)w(t)dt, \int_{t_1}^{r} \Lambda_2(t)w(t)dt)\rangle = 0.$$

Once again we remind the reader the definition of the vector space $\mathcal{V}(t_1, r)$:

$$\mathcal{V}(t_1, r) = Sp\{ (\int_{t_1}^{r} \Lambda_1(t)w(t)dt, \int_{t_1}^{r} \Lambda_2(t)w(t)dt) \mid w \in U[t_1, r] \}.$$

We obtain:

<u>Lemma 6.2.2</u> If $(c_w(t), y(t))$ is an extended extremal trajectory such that the relation

$$(c_w(t), y(t), w(t)) = (c_u(t), x(t), u(t))$$

holds on $[t_1, r]$, then $(\Delta, \gamma) \in \mathcal{V}(t_1, r)^{\perp}$.

<u>Remark 6.2.1</u> Consider $dim\mathcal{V}(t_1, s)$ as a function of s. Choose $s_1 \le s_2$, the definition of $\mathcal{V}(t_1, r)$ implies that $\mathcal{V}(t_1, s_1) \subset \mathcal{V}(t_1, s_2)$. On the other hand, for each s the space $\mathcal{V}(t_1, s)$ is a subspace of R^{n+k} and, therefore, $dim\mathcal{V}(t_1, s) \le n + k$. This consideration shows that $dim\mathcal{V}(t_1, s)$ is a monotone non decreasing step function with respect to s. This implies that the number of its discontinuity points is finite.

Consider a pair (Δ, γ) and denote $u(t; \eta + \Delta, \mu + \gamma)$ by $w(t)$ and $x(t; \eta + \Delta, \mu + \gamma)$ by $y(t)$. In the next statement we derive conditions which ensure matching of extended extremal extremals.

<u>Lemma 6.2.3</u> If $(\Delta, \gamma) \in \mathcal{V}(t_1, r)^{\perp}$, $\mu + \gamma \in P^{+}$, $|(\Delta, \gamma)| < \epsilon_1$ and $(c_w(t_1), y(t_1)) = (c_u(t_1), x(t_1))$, then the relation $(c_w(t), y(t), w(t)) = (c_u(t), x(t), u(t))$ holds on $[t_1, r]$, and $(c_w(t), y(t))$ is an extended extremal trajectory.

<u>Proof.</u> Denote $\Delta\Phi(t_1, t)$ by $\Delta(t)$, then due to the condition $(\Delta, \gamma) \in \mathcal{V}(t_1, r)^{\perp}$, the relation

$$H_u(t, x(t), u(t), \eta(t) + \Delta(t), \mu + \gamma) = 0 \text{ holds on } [t_1, r].$$

This implies that $(x(t), u(t), \eta(t) + \Delta(t))$ is an extremal triple of $(\mu + \gamma)$−control problem, namely $x(t), \eta(t) + \Delta(t))$ and $(y(t), \xi(t))$ are the solutions of the same $(\mu+\gamma)$−Hamiltonian system (6.1.4), 6.1.5) with the same initial conditions $(y(t_1), \xi(t_1))$. This yields

$$x(t) = y(t), \quad \eta(t) + \Delta(t) = \xi(t) \text{ on } [t_1, r],$$

which implies

$$u(t) = u(t, x(t), \eta(t)) = u(t, x(t), \eta(t) + \Delta(t)) = u(t, y(t), \xi(t)) = w(t) \text{ on } [t_1, r].$$

This completes the proof.

§6.3 Branching of extended extremal trajectories

The results obtained so far concern the necessary and sufficient conditions for matching of the extremal pairs. Namely, the main result of the previous section is as follows: An extremal pair $(x(t; \xi, \lambda), u(t; \xi, \lambda))$ coincides with the extremal pair $(x(t; \eta, \mu), u(t; \eta, \mu))$ if and only if

$$(\xi - \eta, \lambda - \mu) \in \mathcal{V}(t_1, r)^\perp. \tag{6.3.1}$$

In this section we prove that condition (6.3.1) is also a necessary condition for matching of extended extremal trajectories. Denote $x(t; \xi, \lambda)$ by $y(t)$, $u(t; \xi, \lambda)$ by $w(t)$, $x(t; \eta, \mu)$ by $x(t)$ and $u(t; \eta, \mu)$ by $u(t)$. We intend to show that the conditions:

1. $(c_w(t_1), y(t_1)) = (c_u(t_1), x(t_1))$,
2. $(c_w(r), y(r)) = (c_u(r), x(r))$

imply that $(\xi - \eta, \lambda - \mu) \in \mathcal{V}(t_1, r)^\perp$. In other words, the neighboring extended extremal trajectories initiating at the same point and intersecting each other at the time r must coincide on the initial time subinterval $[t_1, r]$. Conditions $1 - 2$ above imply that

$$0 = \int_{t_1}^{r} \{ f_\mu(t, y(t), w(t)) - f_\mu(t, x(t), u(t)) \} dt.$$

A straight computation shows that in this case, namely where $y(t_1) = x(t_1)$ and $y(r) = x(r)$

$$\int_{t_1}^{r} \{ f_\mu(t, y(t), w(t)) - f_\mu(t, x(t), u(t)) \} dt =$$

$$= \int_{t_1}^{r} \begin{pmatrix} y(t) - x(t) \\ w(t) - u(t) \end{pmatrix}^* \begin{pmatrix} H_{xx}(t, x(t), u(t), \eta(t), \mu), & H_{ux}(t, x(t), u(t), \eta(t), \mu) \\ H_{xu}(t, x(t), u(t), \eta(t), \mu), & H_{uu}(t, x(t), u(t), \eta(t), \mu) \end{pmatrix} \begin{pmatrix} y(t) - x(t) \\ w(t) - u(t) \end{pmatrix} dt +$$

$$+ \int_{t_1}^{r} o(\ |(y(t) - x(t), w(t) - u(t))|^2\ ) dt.$$

(See [11] p. 357.) Hence, there exists a positive scalar ϵ_2, such that the inequality $|(\xi - \eta, \lambda - \mu)| < \epsilon_2$ implies that the relations

$$y(t) = x(t), \quad w(t) = u(t) \text{ hold almost everywhere on } [t_1, r].$$

Then, due to Lemma 6.2.2, $(\xi - \eta, \lambda - \mu) \in \mathcal{V}(t_1, r)^{\perp}$.

We are able now to present the definition of ϵ^{**}, as follows:

Definition 6.3.1 Let ϵ^{**} be $\min(\epsilon_1, \epsilon_2)$. This ϵ^{**} determines the set of the neighboring extended extremal trajectories as follows: An extended extremal trajectory $(c_w(t), y(t))$ is a neighboring extended extremal trajectory if

$$\max_{t_1 \le t \le t_2} |(c_u(t), x(t), \eta(t), \mu) - (c_w(t), y(t), \xi(t), \lambda)| < \epsilon^{**}.$$

Theorem 6.3.1 Consider a neighboring extended extremal trajectory $(x_w(t), y(t))$ generated by (ξ, λ). The relation

$$(c_w(t), y(t)) = (c_u(t), x(t))$$

holds on $[t_1, r]$ if and only if $(\xi - \eta, \lambda - \mu) \in \mathcal{V}(t_1, r)^{\perp}$.

Proof. The "if" part follows from the above considerations, the "only if" part follows from Lemma 6.2.2.

We are now in a position to state and prove the main results of this section, namely the necessary and sufficient conditions for branching of extended extremal trajectories.

Theorem 6.3.2 Consider a neighboring extended extremal trajectory $(x_w(t), y(t))$ generated by (ξ, λ). The two following conditions are equivalent:

1. $(c_w(t), y(t)) = (c_u(t), x(t))$ on $[t_1, r]$ and $(c_w(t), y(t)) \neq (c_u(t), x(t))$ on $(r, t_2]$;
2. $(\xi - \eta, \lambda - \mu) \in \mathcal{V}(t_1, r)^{\perp}$ and $(\xi - \eta, \lambda - \mu) \notin \mathcal{V}(t_1, r^+)^{\perp}$.

Proof. Suppose that the first condition holds, then due to Theorem 6.3.1,

$$(\xi - \eta, \lambda - \mu) \in \mathcal{V}(t_1, r)^{\perp}.$$

If for some $\epsilon > 0$

$$(\xi - \eta, \lambda - \mu) \in \mathcal{V}(t_1, r + \epsilon)^{\perp},$$

then by using Lemma 6.2.2 we obtain $(c_w(t), y(t)) = (c_u(t), x(t))$ on $[t_1, r + \epsilon]$–a contradiction to the first condition.

Suppose now that the second condition holds, then Lemma 6.2.2 immediately implies that $(c_w(t), y(t)) = (c_u(t), x(t))$ on $[t_1, r]$. If for some $\epsilon > 0$

$$(c_w(t), y(t)) = (c_u(t), x(t)) \text{ on } [t_1, r + \epsilon],$$

then, using Theorem 6.3.1, we obtain that $(\xi - \eta, \lambda - \mu) \in \mathcal{V}(t_1, r + \epsilon)^{\perp}$–a contradiction to the second condition. The theorem is therefore proved.

§6.4 Branching pairs

In this section we investigate the set of branching pairs of the extended extremal trajectory $(c_u(t), x(t))$ which is defined on the time interval $[t_1, t_2]$. Namely we are looking for the pairs $\{l, r\}$ such that there exists a neighboring extended extremal trajectory $(c_w(t), y(t))$, and the following conditions hold:

1. $(c_w(t), y(t)) = (c_u(t), x(t))$ on $[l, r]$,
2. $(c_w(t), y(t)) \neq (c_u(t), x(t))$ for $t \in [t_1, l) \bigcup (r, t_2]$.

First we wish to consider a special particular case, namely we are interested in the branching pairs of the form $\{l, t_2\}$. It turns out that we actually looking for the points l such, that there exists an extended extremal trajectory which differs from $(c_u(t), x(t))$ on the initial subinterval $[t_1, l)$, and coincides with $(c_u(t), x(t))$ on $[l, t_2]$. We shall say that the point l is a left branching point of the extended extremal trajectory $(c_u(t), x(t))$.

The structure of the branching in this case is absolutely symmetric to the structure of the branching which was considered in the preceding section. For this reason we wish only to display the necessary definitions and to state the main theorems without proofs.

<u>Definition 6.4.1</u> For each real number l define the vector space $\mathcal{V}(t_2, l^-)$ as follows:

$$\mathcal{V}(t_2, l^-) = \bigcap_{h>0} \mathcal{V}(t_2, l - h).$$

<u>Theorem 6.4.1</u> Consider a neighboring extended extremal trajectory $(x_w(t), y(t))$ generated by (ξ, λ). The two following conditions are equivalent:

1. $(c_w(t), y(t)) = (c_u(t), x(t))$ on $[l, t_2]$ and $(c_w(t), y(t)) \neq (c_u(t), x(t))$ on $[t_1, l)$,
2. $(\xi - \eta, \lambda - \mu) \in \mathcal{V}(t_2, l)^\perp$ and $(\xi - \eta, \lambda - \mu) \notin \mathcal{V}(t_2, l^-)^\perp$.

Now we present the main result of the section, namely the necessary and sufficient conditions for existence of the branching pairs.

<u>Theorem 6.4.2</u> A pair $\{l, r\}$ is a branching pair of an extended extremal trajectory $(c_u(t), x(t))$ if and only if the following conditions hold:

1. $\mathcal{V}(l, r^+)^\perp$ is a proper subspace of $\mathcal{V}(l, r)^\perp$,
2. $\mathcal{V}(r, l^-)^\perp$ is a proper subspace of $\mathcal{V}(r, l)^\perp$.

<u>Proof.</u> The proof follows from Theorem 6.3.2, Theorem 6.4.1 and Theorem 5.3.5.

Chapter 7

Optimal control problems with constraints

7.1 Introduction

In this chapter we consider the structure of branching of extremals in the optimal control problem (1.3.1), (1.3.2) with constraints. We deal with two following cases:

1. Smooth constraints, namely constraints which are defined by the equations of the type

$$G_i(t, x(t), u(t)) = 0. \tag{7.1.1}$$

2. Inequality constraints, i.e. constraints which are defined by the inequalities

$$G_i(t, x(t), u(t)) \leq 0. \tag{7.1.2}$$

In both cases $G_i(t, x, u)$ is assumed to be differentiable with respect to (x, u) and measurable in t.

The main mathematical tool which will be used in the chapter is the Pontryagin Maximum Principle. Let $\Omega(t, x)$ be the set of all u which satisfies the inequality constraints for a given pair (t, x), i.e.

$$\Omega(t, x) = \{ \, u \mid u \in R^m, \ G_1(t, x, u) \leq 0, ..., G_k(t, x, u) \leq 0 \, \}$$

We wish to emphasize that in the case of the inequality constraints the control function $u(t, x, \eta)$ is usually not differentiable with respect to (x, η). For this reason the following modification of assumption 1.5.1 is accepted in this chapter for the optimal control problem (1.3.1), (1.3.2) with the inequality constraints (7.1.2):

Assumption 7.1.1 For each triple (t, x, η) close to $(t, x(t), \eta(t))$ there exists a unique $u(t, x, \eta)$ which is Lipschitz with respect to (x, η) and measurable in t such that the following relation holds on $[t_1, t_2]$:

$$-f(t, x, u(t, x, \eta)) + \eta F(t, x, u(t, x, \eta)) = \max_{u \in \Omega(t,x)} \{ -f(t, x, u) + \eta F(t, x, u) \}. \tag{7.1.3}$$

This assumption still enables one to reduce the study of extremals initiating at the same point x_1 to the study of solutions of the Hamiltonian system:

$$\dot{x}(t) = H_\eta(t, x(t), u(t), \eta(t)) \tag{7.1.4}$$

$$\dot{\eta}(t) = -H_x(t, x(t), u(t), \eta(t)).\tag{7.1.5}$$

Our main goals in this chapter are the following: For a given extremal trajectory $x(t)$:

1. Characterize the set of such points r, that there exists a neighboring extremal $y(t)$ which branches out of $x(t)$ at r.

2. Determine the conditions under which the branching points of $x(t)$ exist or do not exist.

In the case of the control problem with smooth constraints we reduce our problem to the searching of branching points of extremals in a control problem without constraints. The structure of branching in the second case is appreciably different from that of the former one. In particular, as it was shown by Example 1.10.5, the "continuous" branching is possible in this situation.

This chapter is organized as follows: Sections two and three are devoted to the description of branching points in the optimal control problem with smooth constraints. In section two we derive sufficient conditions for matching of extremal trajectories. In section three a necessary condition for matching of extremal trajectories is presented. In the last section we derive necessary and sufficient conditions for branching of extremal trajectories in optimal control problems with inequality constraints.

§7.2 Sufficient condition for matching of extremals

As it was already shown in section 1.7, Chapter 1, condition (1.7.6) is the necessary condition for matching of extremal pairs in the optimal control problem (1.3.1), (1.3.2) with constraints (7.1.1). In this section we continue to derive the necessary condition for matching of extremal trajectories. Let $(x(t), u(t), \eta(t))$ be an extremal triple. By using the matrices $\Phi(t_1, s)$, $B(t)$, $C(t)$ and the space $U_c[t_1, s]$ which have been defined in section 1.7, we introduce the following useful definition.

Definition 7.2.1 For each real number r we define the vector spaces $V_c(t_1, r)$ and $V_c(t_1, r^+)$ by the following formulas:

$$V_c(t_1, r) = Sp\{ \int_{t_1}^{r} \Phi(t_1, t)B(t)C(t)v(t)dt \mid v \in U_c[t_1, r] \},$$

$$V_c(t_1, r^+) = \bigcap_{h>0} V_c(t_1, r + h).$$

The relation (1.7.6) and Definition 7.2.1 yield the following lemma:

Lemma 7.2.1 If $(y(t), w(t), \xi(t))$ is an extremal triple such that $(y(t), w(t)) = (x(t), u(t))$ on $[t_1, r]$, then $\xi(t_1) - \eta(t_1) \in V_c(t_1, r)^\perp$.

Lemma 7.2.2 Suppose that $\Delta(t_1) \in V_c(t_1, r)^\perp$, then $(x(t), u(t), \eta(t) + \Delta(t_1)\Phi(t_1, t))$ is an extremal triple on $[t_1, r]$.

Proof. Denote $\Delta(t_1)\Phi(t_1, t)$ as $\Delta(t)$ and first show that

$$\frac{\partial}{\partial u}\{-f(t, x(t), u(t)) + (\eta(t) + \Delta(t))F(t, x(t), u(t))\} \in Sp\{\frac{\partial G}{\partial u}(t, x(t), u(t))\}.$$

Indeed, Assumption (7.1.1) implies that

$$\frac{\partial}{\partial u}\{-f(t, x(t), u(t)) + \eta(t)F(t, x(t), u(t))\} \in Sp\{\frac{\partial G}{\partial u}(t, x(t), u(t))\}.$$

On the other hand definition of $\Delta(t)$ yields

$$\Delta(t)B(t) \in Sp\{\frac{\partial G}{\partial u}(t, x(t), u(t))\}.$$

We wish now to show that the triple $(x(t), u(t), \eta(t) + \Delta(t))$ satisfies the Pontryagin Maximum Principle, i.e.

$$H(t, x(t), u(t), \eta(t) + \Delta(t)) = \max_{\substack{u \\ G(t, x(t), u) = 0}} H(t, x(t), u, \eta(t) + \Delta(t)) \text{ on } [t_1, t_2].$$

In accordance with Assumption 7.1.1 there exists the control function

$$w(t) = u(t, x(t), \eta(t) + \Delta(t))$$

such that

$$H(t, x(t), w(t), \eta(t) + \Delta(t)) = \max_{\substack{u \\ G(t, x(t), u) = 0}} H(t, x(t), u, \eta(t) + \Delta(t)) \text{ on } [t_1, t_2].$$

Namely

$$H_u(t, x(t), w(t), \eta(t) + \Delta(t)) \in Sp\{\frac{\partial G}{\partial u}(t, x(t), u(t))\}.$$

On the other hand

$$H_u(t, x(t), u(t), \eta(t) + \Delta(t)) \in Sp\{\frac{\partial G}{\partial u}(t, x(t), u(t))\}.$$

This implies that

$$H_{uu}(t, x(t), u(t), \eta(t) + \Delta(t))(w(t) - u(t)) \in Sp\{\frac{\partial G}{\partial u}(t, x(t), u(t))\}.$$

Using the relation

$$w(t) - u(t) \in Sp\{\frac{\partial G}{\partial u}(t, x(t), u(t))\}^\perp,$$

we obtain

$$(w(t) - u(t))^* H_{uu}(t, x(t), u(t), \eta(t) + \Delta(t))(w(t) - u(t)) = 0.$$

In accordance with Hypothesis 1.3.1 the matrix $H_{uu}(t, x(t), u(t), \eta(t) + \Delta(t))$ is negative definite, then, finally, $w(t) = u(t)$. This completes the proof.

§7.3 Necessary condition

In this section we intend to derive necessary conditions for matching of extremals. In the first part of the section we consider a linear optimal control system and the last part is devoted to the nonlinear case.

Linear system with quadratic cost and linear constraints.

In this paragraph we assume that $F(t, x, u)$ is linear in x and u, $G(t, x, u)$ is linear in u and does not depends on x, and $f(t, x, u)$ is quadratic of the type (2.4.1), namely

$$F(t, x, u) = A(t)x + B(t)u, \quad G_i(t, x, u) = g_i(t)u \text{ all } i, \tag{7.3.1}$$

$$f(t, x, u) = \begin{pmatrix} x \\ u \end{pmatrix}^* \begin{pmatrix} R_{11}(t) & R_{12}(t) \\ R_{21}(t) & R_{22}(t) \end{pmatrix} \begin{pmatrix} x \\ u \end{pmatrix}. \tag{7.3.2}$$

Consider extremal triples $(x(t), u(t), \eta(t))$ and $(y(t), w(t), \xi(t))$. In this linear case each extremal trajectory is, at the same time, an optimal one. (A proof can be constructed with minor changes from that of Theorem 8.3.1). Hence the relation:

$$x(t_1) = y(t_1) \text{ and } x(r) = y(r)$$

implies that $u(t) = w(t)$ on $[t_1, r]$, and due to Lemma 7.2.1 $\xi(t_1) - \eta(t_1) \in V_c(t_1, r)^{\perp}$.

We obtain, therefore, the following result:

Theorem 7.3.1 The relation $x(t) = y(t)$ holds on $[t_1, r]$ for extremal trajectories $x(t)$ and $y(t)$ if and only if

$$\xi(t_1) - \eta(t_1) \in V_c(t_1, r)^{\perp}.$$

We wish to mention that for each $t \in [t_1, t_2]$ the mapping $\eta \mapsto x(t; \eta)$ is linear with respect to the second argument, namely for each two scalars α, β

$$x(t; \alpha\eta + \beta\mu) = \alpha x(t; \eta) + \beta x(t; \mu).$$

(Here $x(t;\eta)$ is the first n coordinates of a solution $(x(t),\eta(t))$ of $(7.1.4) - (7.1.5)$ with $x(t_1) = x_1$ and $\eta(t_1) = \eta$.) The assertion follows straightforward from the linear structure of equations $(7.3.1)$. The linearity of $x(t;\eta)$ implies the following equation:

$$rank\frac{\partial x(t;\cdot)}{\partial \xi}\Big|_{\eta(t_1)} = dim\, V_c(t_1,t). \qquad (7.3.3)$$

In the next paragraph we will use equation $(7.3.3)$ for description of branching points in nonlinear control systems.

The nonlinear case.

In the rest of the section our main aim is to show that the condition $\Delta \in V_c(t_1,r)^\perp$ is necessary for matching of the extremal trajectories $x(t;\eta)$ and $x(t;\eta + \Delta)$ on the initial subinterval $[t_1,r]$. Consider a solution $(x(t),\eta(t))$ of the system $(7.1.4) - (7.1.5)$ with $x(t_1) = x_1$, $\eta(t_1) = \xi$. Denote this solution as $(x(t;x_1,\xi),\ \eta(t;x_1,\xi))$. Once selected $x(t_1)$ will not be changed and, therefore, a solution can be unambiguously denoted as $(x(t;\xi),\ \eta(t;\xi))$. We wish to assure the smooth (C^1) dependence of solutions $\{x(t;\xi)\}$ on initial conditions $\{\xi\}$. To this end the additional assumption is introduced.

Assumption 7.3.1 There exists Lebesgue integrable function $m(t)$ such that for each extremal triple $(y(t),w(t),\mu(t))$ generated by $\mu(t_1)$ close to $\eta(t_1)$ the following condition holds:

$$\left|\begin{pmatrix} H_\eta(t,y(t),w(t),\mu(t)) \\ -H_x(t,y(t),w(t),\mu(t)) \end{pmatrix}\right| + \left|\begin{pmatrix} H_{\eta x}(t,y(t),w(t),\mu(t)), & H_{\eta\eta}(t,y(t),w(t),\mu(t)) \\ -H_{xx}(t,y(t),w(t),\mu(t)), & -H_{x\eta}(t,y(t),w(t),\mu(t)) \end{pmatrix}\right| \le m(t).$$

A mapping $x(t;\cdot) : R^n \mapsto R^n$ ($\xi \mapsto x(t;\xi)$) is, therefore, continuously differentiable with respect to ξ. (A proof can be constructed with minor changes from that of Theorem 7.2, p. 25, Coddington and Levinson [3].)

The remainder of the section is devoted to the following question: "how many neighboring extremals intersect $x(s;\eta(t_1))$ at time t?" The answer to this question is closely related to knowledge of $rank\ \frac{\partial x(t;\cdot)}{\partial \xi}\big|_{\eta(t_1)}$, where by $\frac{\partial x(t;\cdot)}{\partial \xi}\big|_{\eta(t_1)}$ we denote the partial derivative of $x(t;\xi)$ with respect to the second argument evaluated at the point $(t;\eta(t_1))$.

In order to answer the question we intend to construct a linear control system with linear constraints of the form $(7.3.1)$ in such a way that

$$rank\frac{\partial x(t;\cdot)}{\partial \xi}\Big|_{\eta(t_1)} = rank\frac{\partial z(t;\cdot)}{\partial \theta}\Big|_0 \quad \text{for each } t \in [t_1,t_2].$$

Here $x(t)$ is the extremal trajectory of the control problem $(1.3.1)$, $(1.3.2)$ with the smooth constraints $(7.1.1)$, and $z(t)$ is an extremal trajectory of the linear control problem $(7.3.1)$, $(7.3.2)$.

The matrices $A(t)$ and $B(t)$ had been already defined and all that we have left to construct is a cost functional $h(t, z, v)$ and the linear constraints $(g_1(t), ..., g_k(t))$. Define $h(t, z, v)$ and the constraints $(g_1(t), ..., g_k(t))$ as follows:

$$h(t, z, v) = -\frac{1}{2} \begin{pmatrix} z \\ v \end{pmatrix}^* \begin{pmatrix} H_{xx}(t, x(t), u(t), \eta(t)) & H_{ux}(t, x(t), u(t), \eta(t)) \\ H_{xu}(t, x(t), u(t), \eta(t)) & H_{uu}(t, x(t), u(t), \eta(t)) \end{pmatrix} \begin{pmatrix} z \\ v \end{pmatrix},$$

$$g_i(t) = \frac{\partial G_i}{\partial u}(t, x(t), u(t)), \quad i = 1, .., k; \text{ where by } * \text{ we indicate the transpose.}$$

In this work we consider first order necessary and sufficient conditions for branching. Namely, we investigate properties of nonlinear control problems by studying their linearizations, or first order estimations (see Chapter 2, section 2.4). In order to be able to apply this technique we need to make the additional assumption:

<u>Assumption 7.3.2</u> The vectors $\{g_i(t)\}$, $i = 1, .., k$, are linearly independent for $t \in [t_1, t_2]$.

Denote by a $(n + n) \times (n + n)$ matrix

$$\Psi(t, t_1) = \begin{pmatrix} \Psi_{11}(t, t_1) & \Psi_{12}(t, t_1) \\ \Psi_{21}(t, t_1) & \Psi_{22}(t, t_1) \end{pmatrix}, \quad \Psi(t_1, t_1) = I.$$

the fundamental solution of the following linear differential equation:

$$\dot{X} = \begin{pmatrix} H_{\eta x}(t, x(t), u(t), \eta(t)) & H_{\eta\eta}(t, x(t), u(t), \eta(t)) \\ -H_{xx}(t, x(t), u(t), \eta(t)) & -H_{x\eta}(t, x(t), u(t), \eta(t)) \end{pmatrix} X.$$

Here $X \in R^{n+n}$. In this case

$$\frac{\partial x(t; \cdot)}{\partial \xi}\Big|_{\eta(t_1)} = \Psi_{12}(t, t_1) \tag{7.3.4}$$

(see [3], p. 25). On the other hand a direct computation shows that

$$\frac{\partial z(t; \cdot)}{\partial \theta}\Big|_0 = \Psi_{12}(t, t_1) \text{ and, therefore } \frac{\partial z(t; \cdot)}{\partial \theta}\Big|_0 = \frac{\partial x(t; \cdot)}{\partial \xi}\Big|_{\eta(t_1)}.$$

This implies that $rank \frac{\partial z(t; \cdot)}{\partial \xi}\Big|_{\eta(t_1)} = dim\, V_c(t_1, t)$ (see equation (7.3.3)). Hence, due to the Implicit Function Theorem, For each $t \in [t_1, t_2]$ there exists $\epsilon(t) > 0$ such that the set

$$\Xi(t) = \{ \xi \mid \xi \in R^n, \ |\xi - \eta(t)| < \epsilon(t), \ x(t; \xi) = x(t; \eta(t_1)) \}$$

is a subset of a $n\text{-}dimV_c(t_1, t)$ dimensional manifold. On the other hand we already know that for each Δ small enough, such that $\Delta \in V_c(t_1, r)^\perp$ the relation $x(t; \eta) = x(t; \eta + \Delta)$ holds on $[t_1, r]$. This implies that the condition

$$\Delta \in V_c(t_1, r)^\perp$$

is the necessary and sufficient condition for the matching of extremals.

We introduce now an additional definition and present the main result of this section, namely the necessary and sufficient conditions for branching of the extremal trajectory $x(t; \eta)$.

Definition 7.3.1 We denote the set of discontinuity points of the function $dimV_c(t_1, s)$ on the interval $[t_1, t_2]$ as $\{r_i\}$.

Theorem 7.3.2 Each $r \in \{r_i\}$ is a branching point of an extremal trajectory $x(t)$. Let $y(t)$ be a neighboring extremal trajectory. If $y(t_1) = x(t_1)$ and $y(s) = x(s)$ for some $s \in [t_1, t_2]$, then there exists $r \in [t_1, s]$ such that the following conditions hold:

1. $r \in \{r_i\}$,
2. $y(t) = x(t)$ on $[t_1, r]$,
3. $y(t) \neq x(t)$ on $(r, t_2]$.

Corollary 7.3.1 The set of branching points of $x(t)$ coincides with that of the following linear control system without constraints:

$$\dot{z}(t) = A(t)z(t) + B(t)C(t)v(t).$$

Corollary 7.3.2 The set of branching points of an extremal trajectory $x(t)$ is a finite set. The number of its elements is no more than the dimension of the state space R^n.

Further extensions.

As the next step we consider the set of branching pairs of the extremal trajectory $x(t)$. Namely, we consider the pairs $\{l, r\}$, $t_1 \leq l < r \leq t_2$ such that for each $\{l, r\}$ there exists a neighboring extremal trajectory $y(t)$ which branches out of $x(t)$ at $\{l, r\}$, i.e. $x(t) = y(t)$ over $[r, l]$, and $x(t) \neq y(t)$ for $t \in [t_1, l) \cup (r, t_2]$.

First we consider the special particular case, namely the pairs $\{l, r\}$ such that $r = t_2$, i.e. actually we are interested in points l, such that there exists an extremal trajectory $y(t)$ which is different from $x(t)$ on $[t_1, l)$, and coincides with $x(t)$ on $[l, t_2]$. We call such point l a left branching point of the extremal trajectory $x(t)$. In this case one can easily prove the following modification of Corollary 7.3.1:

Theorem 7.3.3 The set of branching points of $x(t)$ coincides with that of the following linear control system without constraints:

$$\dot{z}(t) = A(t)z(t) + B(t)C(t)v(t).$$

The main result in the case of branching pairs is as follows:

Theorem 7.3.4 The set of branching pairs of an extremal trajectory $x(t)$ coincides with one of the following linear control system without constraints

$$\dot{z} = A(t)z + B(t)C(t)v(t).$$

Here $A(t) = \frac{\partial F}{\partial x}(t, x(t), u(t))$, $B(t) = \frac{\partial F}{\partial u}(t, x(t), u(t))$ and $C(t)$ is an orthogonal projection of R^m on $Sp\{ \frac{\partial G}{\partial u}(t, x(t), u(t)) \}^\perp$.

Proof. The proof follows from Corollary 7.3.1 and Theorem 7.3.3.

§7.4 Nonlinear case with inequality constraints

In this section we examine extremals of the optimal control problem (1.3.1), (1.3.2) with inequality constraints (7.1.2). Our main purpose is to describe the set of branching points (as well as branching pairs) of the extremal trajectory $x(t)$. We derive first, the necessary condition for matching of the extremal trajectories. Denote by $co(t)$ the following closed, convex cone

$$\{ z \mid z \in R^m, \ z = \sum_{i=1}^{k} \alpha_i \frac{\partial G_i}{\partial u}(t, x(t), u(t)), \ \alpha_i \geq 0 \text{ for all } i, \ \alpha_j = 0 \text{ if } G_j(t, x(t), u(t)) < 0. \}$$

In accordance with the Pontryagin Maximum Principle the relation

$$H_u(t, x(t), u(t), \eta(t)) \in co(t) \text{ holds on } [t_1, t_2]. \tag{7.4.1}$$

Suppose, that $(y(t), w(t), \xi(t))$ is an extremal triple such that $(x(t), u(t)) = (y(t), w(t))$ on $[t_1, r]$. This implies that

$$H_u(t, y(t), w(t), \xi(t)) \in co(t) \text{ on } [t_1, r].$$

In other words

$$H_u(t, x(t), u(t), \eta(t)) + (\xi(t_1) - \eta(t_1))\Phi(t_1, t)B(t) \in co(t) \text{ on } [t_1, r].$$

Denote by $C(t_1, s)$ the following convex set

$$C(t_1, s) = \{ \Delta \mid \Delta \in R^n, \ H_u(t, x(t), u(t), \eta(t)) + \Delta\Phi(t_1, t)B(t) \in co(t), \ t \in [t_1, s]. \}$$

We wish to emphasize two evident, but extremely useful properties of $C(t_1, s)$:

1. $C(t_1, s)$ is a convex, closed set for each $s \in [t_1, t_2]$,
2. $C(t_1, s_1) \supset C(t_1, s_2)$ provided $s_1 \leq s_2$.

In the next statements we derive necessary and sufficient conditions for branching of $x(t)$.

Theorem 7.4.1 If $(x(t), u(t)) = (y(t), w(t))$ on $[t_1, r]$, then $\xi(t_1) - \eta(t_1) \in C(t_1, r)$.

Proof. First, note that $H_u(t, x(t), u(t), \xi(t)) \in co(t)$ on $[t_1, r]$. Next, by using the relation

$$H_u(t, x(t), u(t), \xi(t)) = H_u(t, x(t), u(t), \eta(t)) + (\xi(t_1) - \eta(t_1))\Phi(t_1, t)B(t)$$

and the definition of $C(t_1, r)$ we obtain the desired result.

Theorem 7.4.2 There exists a positive ϵ^{***} such that for each vector $\Delta \in C(t_1, r)$ with $|\Delta| < \epsilon^{***}$ the triple $(x(t), u(t), \eta(t) + \Delta\Phi(t_1, t))$ is an extremal triple of the optimal control problem (1.3.1), (1.3.2), (7.1.2) on $[t_1, r]$.

We wish to mention, that in accordance with Hypothesis 1.3.1 the matrix $H_{uu}(t, x(t), u(t), \eta(t))$ is negative definite on $[t_1, t_2]$. Hence, if $|\Delta|$ is small, $H_{uu}(t, x(t), u(t), \eta(t) + \Delta\Phi(t_1, t))$ is also negative definite on $[t_1, t_2]$. This observation will be useful in the following proof.

Proof. All we have to prove is that

$$-f(t, x(t), u(t)) + \{\eta(t) + \Delta\Phi(t_1, t)\}F(t, x(t), u(t)) =$$

$$= \max_{\substack{u \\ G(t, x(t), u) \le 0}} \{-f(t, x(t), u) + \{\eta(t) + \Delta\Phi(t_1, t)\}F(t, x(t), u)\} \text{ on } [t_1, t_2].$$

Indeed, suppose the opposite, namely that there exists an admissible control $w(t)$ which is different from $u(t)$ on $[t_1, r]$ and

$$-f(t, x(t), w(t)) + \{\eta(t) + \Delta\Phi(t_1, t)\}F(t, x(t), w(t)) =$$

$$= \max_{\substack{u \\ G(t, x(t), u) \le 0}} \{-f(t, x(t), u) + \{\eta(t) + \Delta\Phi(t_1, t)\}F(t, x(t), u)\} \text{ on } [t_1, t_2].$$

In accordance with equation (7.4.1) $H_u(t, x(t), u(t), \eta(t)) \in co(t)$ on $[t_1, t_2]$, on the other hand the definition of the vector Δ implies, that $H_u(t, x(t), u(t), \xi(t)) \in co(t)$ on $[t_1, r]$, where $\xi(t) = \eta(t) + \Delta\Phi(t_1, t)$. This yields, due to the first condition of Hypothesis 1.3.1, that

$$H_u(t, x(t), u(t), \xi(t))(w(t) - u(t)) \le 0 \text{ provided } |w(t) - u(t)| \text{ is small.}$$

On the other hand $H_u(t, x(t), w(t), \xi(t)) \in co(t)$ and then

$$H_u(t, x(t), w(t), \xi(t))(w(t) - u(t)) \le 0.$$

The two last relations yield the following:

$$0 \ge H_u(t, x(t), w(t), \xi(t))(w(t) - u(t)) = H_u(t, x(t), u(t), \xi(t))(w(t) - u(t)) +$$

$$+ (w(t) - u(t))^* H_{uu}(t, x(t), u(t), \xi(t))(w(t) - u(t)) + o(|w(t) - u(t)|^2) =$$

$$= H_u(t, x(t), u(t), \xi(t))(w(t) - u(t)) + o(|w(t) - u(t)|) \ge 0.$$

This implies that $H_u(t, x(t), u(t), \xi(t))(w(t) - u(t)) = 0$. Hence,

$$(w(t) - u(t))^* H_{uu}(t, x(t), u(t), \xi(t))(w(t) - u(t)) + o(|w(t) - u(t)|^2) = 0.$$

Namely,

$$(w(t) - u(t))^* H_{uu}(t, x(t), u(t), \xi(t))(w(t) - u(t)) = 0.$$

In accordance with Hypothesis 1.3.1, there exist positive constants m and δ such that for each vector $v \in R^n$ the condition

$$|(t, x(t), u(t), \eta(t)) - (t, y, w, \xi)| < \delta$$

implies that

$$v^* H_{uu}(t, y, w, \xi)v \leq -\frac{m}{2}|v|^2.$$

Choose now a positive ϵ^{***} in such a way, that $|\Delta\Phi(t_1, t)| < \delta$ for each $t \in [t_1, t_2]$ provided $|\Delta| < \epsilon^{***}$. Then, we obtain

$$0 = (w(t) - u(t))^* H_{uu}(t, x(t), u(t), \xi(t))(w(t) - u(t)) =$$

$$= (w(t) - u(t))^* H_{uu}(t, x(t), u(t), \eta(t) + \Delta\Phi(t_1, t))(w(t) - u(t)) \leq -\frac{m}{2}|w(t) - u(t)|^2.$$

This implies that $w(t) = u(t)$ and completes the proof.

From now on this ϵ^{***} define the family of neighboring extremals, as follows: an extremal $y(t)$ is a neighboring extremal trajectory if

$$|(x(t), \eta(t)) - (y(t), \xi(t))| < \epsilon^{***} \text{ on } [t_1, t_2].$$

<u>Corollary 7.4.1</u> If $y(t)$ is a neighboring extremal trajectory such that $\xi(t_1) - \eta(t_1) \in C(t_1, r)$, then

$$(x(t), u(t)) = (y(t), w(t)) \text{ on } [t_1, r].$$

<u>Proof.</u> The vectors $(x(t), \eta(t) + \Delta\Phi(t_1, t))$ and $(y(t), \xi(t))$ are solutions of the Hamiltonian system on $[t_1, r]$. Note, that $(x(t_1), \eta(t_1) + \Delta\Phi(t_1, t_1)) = (y(t_1), \xi(t_1))$. The proof now follows from Theorem 7.4.2.

<u>Theorem 7.4.3</u> Consider a neighboring extremal trajectory $y(t)$. Denote $\xi(t_1) - \eta(t_1)$ by Δ. $x(t_1) = y(t_1)$ and $x(r) = y(r)$ for some $r \in [t_1, t_2]$, then the following relations hold:

1. $(x(t), u(t)) = (y(t), w(t))$ on $[t_1, r]$,
2. $\Delta \in C(t_1, r)$.

<u>Proof.</u> Note that in this case, namely where $x(t_1) = y(t_1)$ and $x(r) = y(r)$

$$\int_{t_1}^r \{ f(t, y(t), w(t)) - f(t, x(t), u(t)) \} dt =$$

$$= \int_{t_1}^r \begin{pmatrix} y(t) - x(t) \\ w(t) - u(t) \end{pmatrix}^* \begin{pmatrix} H_{xx}(t, x(t), u(t), \eta(t)) & H_{ux}(t, x(t), u(t), \eta(t)) \\ H_{xu}(t, x(t), u(t), \eta(t)) & H_{uu}(t, x(t), u(t), \eta(t)) \end{pmatrix} \begin{pmatrix} y(t) - x(t) \\ w(t) - u(t) \end{pmatrix} dt +$$

$$+ \int_{t_1}^{r} o(\, |(y(t) - x(t), w(t) - u(t))|^2 \,)dt > 0$$

and

$$\int_{t_1}^{r} \{ \, f(t, x(t), u(t)) - f(t, y(t), w(t)) \, \}dt =$$

$$= \int_{t_1}^{r} \begin{pmatrix} y(t) - x(t) \\ w(t) - u(t) \end{pmatrix}^{*} \begin{pmatrix} H_{xx}(t, y(t), w(t), \xi(t)) & H_{ux}(t, y(t), w(t), \xi(t)) \\ H_{xu}(t, y(t), w(t), \xi(t)) & H_{uu}(t, y(t), w(t), \xi(t)) \end{pmatrix} \begin{pmatrix} y(t) - x(t) \\ w(t) - u(t) \end{pmatrix} dt +$$

$$+ \int_{t_1}^{r} o(\, |(y(t) - x(t), w(t) - u(t))|^2 \,)dt > 0.$$

(See [11], p. 357.) This implies, due to the second condition of Hypothesis 1.3.1, that $(x(t), u(t)) = (y(t), w(t))$. Hence, in accordance with Theorem 7.4.1, $\Delta \in C(t_1, r)$.

Theorem 7.4.2, Theorem 7.4.3 and Definition 1.7.2 yield the main result of the paragraph, namely:

Theorem 7.4.4 A point r is a branching point of the extremal trajectory $x(t)$ if and only if $C(t_1, r) \neq C(t_1, r^+)$.

The two sided branching.

In the present paragraph we examine the structure of two sided branching of an extremal trajectory $x(t)$. The main result obtained in this paragraph, namely the necessary and sufficient conditions for branching, is a natural generalization of Theorem 7.3.4.

First we introduce the necessary definitions.

Definition 7.4.1 For each pair of real numbers l, r we define the following convex sets:

$$C(l, r) = \{ \, \Delta \mid \Delta \in R^n, \; H_u(t, x(t), u(t), \eta(t)) + \Delta\Phi(t_1, t)B(t) \in co(t), \; t \in [l, r] \, \}$$

$$C(l, r^+) = \bigcup_{h > 0} C(l, r + h),$$

$$C(r, l^-) = \bigcup_{h > 0} C(r, l - h).$$

Remark 7.4.1 In this section the role of $C(l, r)$ is similar to that of $V_c(l, r)^{\perp}$ in the problems with smooth constraints. Note, that the following relation holds for $l, r \in [t_1, t_2]$

$$C(l, r) = C(r, l)\Phi(l, r).$$

Theorem 7.4.5 A pair $\{l, r\}$ is a branching pair of the extremal trajectory $x(t)$ if and only if there exists $\Delta \in R^n$ such that the following conditions hold:

1. $\Delta \in C(l,r)$,
2. $\Delta \Phi(r,l) \in C(r,l)$,
3. $\Delta \notin C(l,r^+)$,
4. $\Delta \Phi(r,l) \notin C(r,l^-)$.

Proof. Consider a solution $(y(t), \xi(t))$ of the system (7.1.4), (7.1.5) with $y(r) = x(r)$ and $\xi(r) = \eta(r) + \Delta$. Corollary 7.4.1 implies that $x(t) = y(t)$ on $[l,r]$, then, due to Theorem 7.4.3, $x(t) \neq y(t)$ on $(r, t_2]$. By minor change of Theorem 7.4.3 one can show that $x(t) \neq y(t)$ on $[t_1, l)$.

§8.1 Statement of the problem

Consider a linear control system:

$$\frac{d}{dt}x(t) = A(t)x(t) + B(t)u(t) \tag{8.1.1}$$

where $x \in R^n$, $u \in R^m$ and $A(t)$ and $B(t)$ are matrix valued mappings with the appropriate dimensions.

Consider a cost functional

$$c(x, u) = \int_{t_1}^{t_2} f(t, x(t), u(t))dt. \tag{8.1.2}$$

We assume that the following conditions hold:

1. $A(t)$, $B(t)$ are the $n \times n$ and $n \times m$ matrices whose elements are Lebesgue integrable on $[t_1, t_2]$,

2. $f(t, x, u)$ is twice continuously differentiable with respect to x and u,

3. $f(t, x, u) \geq \alpha |u|^p$ for some $p > 1$, $\alpha > 0$,

4. $f(t, x, u)$ is strictly convex with respect to (x, u),

5. $\frac{\partial^2 f}{\partial u^2}(t, x, u) \geq Q$ for each $(t, x, u) \in R \times R^n \times R^m$ and for some positive definite matrix Q,

    where $\frac{\partial^2 f}{\partial u^2}(t, x, u)$ is the matrix $\{\frac{\partial^2 f}{\partial u_i \partial u_j}\}_{i,j=1,..,m}$.

The main goal of this section is a construction of the Hamiltonian system (8.2.1), (8.2.2) below and one-to-one correspondence between optimal trajectories of the optimal control problem (8.1.1), (8.1.2) and solutions of the system (8.2.1), (8.2.2).

8.2 Construction of the system of differential equations

Condition 5 which has been presented in the preceding section and the Implicit Function Theorem (see [21], p.41) imply that for each (t, x, η) there exists a unique $v(t, x, \eta)$ such that:

$$\frac{\partial f}{\partial u}(t, x, v(t, x, \eta)) = \eta B(t),$$

where $\frac{\partial f}{\partial u} = (\frac{\partial f}{\partial u_1}, ..., \frac{\partial f}{\partial u_m})$.

This allows us to introduce the following definition.

Definition 8.2.1 Let

$$v : R \times R^n \times R^n \mapsto R^m$$

be defined by the equation

$$\frac{\partial f}{\partial u}(t, x, v(t, x, \eta)) = \eta B(t).$$

We wish to recall that $(x(t), u(t))$ is an optimal pair of (8.1.1), (8.1.2) if for any admissible pair $(y(t), w(t))$ with $y(t_1) = x(t_1)$ and $y(t_2) = x(t_2)$ the following condition holds:

$$c(x, u) - c(y, w) \leq 0.$$

In the next lemma we show that each optimal pair $(x(t), u(t))$ of the optimal control problem (8.1.1), (8.1.2) is associated with a certain solution of the system (8.2.1), (8.2.2) presented below.

Lemma 8.2.1 For each optimal pair $(x(t), u(t))$ of the optimal control problem (8.1.1), (8.1.2) there exists a solution $(y(t), \eta(t))$ of the system

$$\frac{d}{dt}y(t) = A(t)y(t) + B(t)v(t, y(t), \eta(t)) \qquad (8.2.1)$$

$$\frac{d}{dt}\eta(t) = \frac{\partial f}{\partial x}(t, y(t), v(t, y(t), \eta(t))) - \eta(t)A(t) \qquad (8.2.2)$$

such that $y(t) = x(t)$ and $v(t, y(t), \eta(t)) = u(t)$.

Proof. The Pontryagin Maximum Principle holds for the optimal pair $(x(t), u(t))$, namely, there exists a vector function $\eta(t)$–a solution of the ordinary differential equation

$$\dot{\eta} = \frac{\partial f}{\partial x}(t, x(t), u(t)) - \eta A(t)$$

such that

$$-f(t, x(t), u(t)) + \eta(t)B(t)u(t) = \max_{w}\{-f(t, x(t), w) + \eta(t)B(t)w\},$$

(see [11], p. 319). Hence, the relation

$$\frac{\partial f}{\partial u}(t, x(t), u(t)) = \eta(t)B(t) \text{ holds on } [t_1, t_2].$$

Namely,

$$u(t) = v(t, x(t), \eta(t))$$

and, therefore,

$$\frac{d}{dt} x(t) = A(t)x(t) + B(t)v(t, x(t), \eta(t))$$

$$\frac{d}{dt} \eta(t) = \frac{\partial f}{\partial x}(t, x(t), \eta(t)) - \eta(t)A(t).$$

We say that $(y(t), \eta(t))$ is a solution of the system (8.1.1), (8.1.2) generated by the optimal pair $(x(t), u(t))$. In the same manner one can produce a trajectory of (8.1.1) from a solution of (8.2.1), (8.2.2). More precisely:

1. Let $(y(t), \eta(t))$ be a solution of the system (8.2.1), (8.2.2).

2. Define $z(t)$, $u(t)$–a trajectory and a corresponding control of (8.1.1) by the following equations:

$$z(t) = y(t), \quad u(t) = v(t, y(t), \eta(t)).$$

We denote the trajectory $z(t)$ produced by $(y(t), \eta(t))$ by $z(t; y(t_1), \eta(t_1))$.

This is the natural place to ask the following question: "For a given solution $(y(t), \eta(t))$ of the system (8.2.1)-(8.2.2), is the created trajectory $z(t; y(t_1), \eta(t_1))$ optimal or not?". The answer will be given at the end of this section (Theorem 8.3.1).

§8.3 Connection between the sets of trajectories

This section is devoted to the investigation of one–to–one correspondence between the set of optimal trajectories of the system (8.1.1) − (8.1.2) and the set of trajectories created by the solutions of the system (8.2.1) − (8.2.2).

We first wish to show that pairs of initial conditions (x_1, η_1), (x_1, η_2) which satisfy the relation $\eta_1 - \eta_2 \in V(t_1, t_2)^\perp$, create the same trajectories of the system (8.2.1), (8.2.2). In other words,

$$z(t; x_1, \eta_1) = z(t; x_1, \eta_2) \quad \text{on} \quad [t_1, t_2].$$

Roughly speaking this relation implies, that it is enough to treat the trajectories created by the initial conditions of the form:

$$(x_1, \eta), \quad \eta \in V(t_1, t_2).$$

More precisely:

Consider a solution $(y(t), \mu(t))$ of the system (8.2.1), (8.2.2). Choose $\xi \in V(t_1, t_2)^\perp$. This choice of ξ implies that

$$\mu(t)B(t) = \mu(t)B(t) + \xi\Phi(t_1, t)B(t).$$

This yields

$$v(t, y(t), \mu(t)) = v(t, y(t), \mu(t) + \xi \Phi(t_1, t)),$$

and, therefore, $(y(t), \mu(t) + \xi \Phi(t_1, t))$ is also a solution of (8.2.1), (8.2.2). We summarize the above discussion by the following statement.

<u>Corollary 8.3.1</u> Let $(y_1(t), \mu_1(t))$ and $(y_2(t), \mu_2(t))$ be solutions of the system (8.2.1), (8.2.2). If

$$y_1(t_1) = y_2(t_1), \quad \text{and} \quad \mu_1(t_1) - \mu_2(t_1) \in V(t_1, t_2)^{\perp},$$

then these solutions generate the same trajectories of (8.1.1), namely,

$$z(t; y_1(t_1), \mu_1(t_1)) = z(t; y_2(t_1), \mu_2(t_1)).$$

<u>Definition 8.3.1</u> For $x_1 \in R^n$ define the set $N(x_1)$ as follows:

$$N(x_1) = \{ \eta \in V(t_1, t_2) \mid z(t; x_1, \eta) \text{ is an optimal trajectory of } (8.1.1) - (8.1.2) \}.$$

Our main goal in the rest of the section is to prove the following relation:

$$N(x_1) = V(t_1, t_2).$$

For each $x_1 \in R^n$ we denote the attainable set at time t_2 for the system (8.1.1) with initial time t_1 and initial state x_1 by $AT(x_1; t_1, t_2)$, i.e.,

$$AT(x_1; t_1, t_2) =$$

$\{ x \in R^n \mid \text{there exists an admissible trajectory } \phi(t) \text{ of } (8.1.1) \text{ such that } \phi(t_1) = x_1, \ \phi(t_2) = x \}.$

<u>Lemma 8.3.1</u> For each $x_1 \in R^n$ the set $N(x_1)$ is an open subset of $V(t_1, t_2)$.

For each $x \in AT(x_1; t_1, t_2)$ consider the optimal pair $(x(t), u(t))$ such that $x(t_1) = x_1$ and $x(t_2) = x$. Define a function $C : AT(x_1; t_1, t_2) \mapsto R$ as follows:

$$C(x) = \int_{t_1}^{t_2} f(t, x(t), u(t)) dt.$$

By a slight modification of Theorem 8 (see [11], p. 209) one can show that $C(x)$ is convex and well defined function on $AT(x_1; t_1, t_2)$.

Our proof consists of the following three steps:

First we show that the convex function $C(x)$ is differentiable.

In the second step we prove that $C(x)$ is continuously differentiable.

In the last step we consider the mapping

$$\varphi \; : \; AT(x_1; t_1, t_2) \longmapsto V(t_1, t_2)$$

which is defined by the relation:

$$z(t_2; x_1, \varphi(x)) = x.$$

(Here $z(t; x_1, \eta)$ is an optimal trajectory created by $\eta \in N(x_1)$ such, that $z(t_1; x_1, \eta) = x_1$; see Definition 8.3.1.)

We shall prove that this mapping is well defined, one-to-one and continuous. Then, due to Invariant Domain Theorem (see [12], p. 50),

$$\{ \varphi(\, AT(x_1; t_1, t_2) \,) \, \} \text{ is an open subset of } V(t_1, t_2).$$

On the other hand, definition of φ implies that

$$\varphi(\, AT(x_1; t_1, t_2) \,) = N(x_1)$$

and this completes the proof.

STEP 1. We wish to show that $C(x)$ has a unique support vector of the form $(-1, \eta_0)$ at each $x_0 \in AT(x_1; t_1, t_2)$. To this end we shall construct an auxiliary differentiable function $C_1(x)$ such that:

1. $C_1(x) \geq C(x)$ for each $x \in AT(x_1; t_1, t_2)$,
2. $C_1(x_0) = C(x_0)$.

Suppose that $(x(t), u(t))$ is an optimal pair such that

$$x(t_1) = x_1, \quad x(t_2) = x_0.$$

Let $\{ \, v_1(t), ..., v_k(t) \, \}$ be a set of controls which produce a basis for the linear space $AT(0; t_1, t_2)$, namely

$$Sp\{e_1, .., e_k\} = AT(0; t_1, t_2), \quad e_i = \int_{t_1}^{t_2} \Phi(t_2, t) B(t) v_i(t) dt, \; i = 1, .., k.$$

Since each $x \in AT(x_1; t_1, t_2)$ has a unique representation as

$$x = x_0 + \sum_{i=1}^{k} \alpha_i \int_{t_1}^{t_2} \Phi(t_2, t) B(t) v_i(t) dt,$$

one can unambiguously define $C_1(x)$ as follows:

$$C_1(x) = \int_{t_1}^{t_2} f(t, y(t), w(t)) dt,$$

where

$$y(t) = x(t) + \sum_{i=1}^{k} \alpha_i \int_{t_1}^{t} \Phi(t, s) B(s) v_i(s) ds \quad \text{and} \quad w(t) = u(t) + \sum \alpha_i v_i(t).$$

It is easy to check that $C_1(x)$ satisfies conditions $1-2$. Suppose now that $(-1, \eta_0)$ is a support vector to $C(x)$ at x_0. This implies that $(-1, \eta_0)$ is also a unique support vector to $C_1(x)$ at x_0. This completes the first step.

STEP 2. We wish to show that $C(x)$ is continuously differentiable. Suppose the opposite, namely, there exists a sequence $\{x_i\}$ in $AT(x_1; t_1, t_2)$ with the corresponding sequence of the support vectors $\{(-1, \eta_i)\}$ such that

1. $x_i \to x_0$,
2. $|\eta_i - \eta_0| > \epsilon$ for some $\epsilon > 0$.

Let $O_\delta = \{ \, x \mid x \in AT(x_1; t_1, t_2), \; |x - x_0| < \delta \, \}$.

Choose $\delta > 0$ such that for each $x \in O_\delta$ the relation

$$|\langle (-1, \eta_0), (C(x) - C(x_0), x - x_0) \rangle| < \frac{\epsilon}{4} |x - x_0| \text{ holds.}$$

Next, note that

$$|\langle (-1, \eta_0), (C(x) - C(x_i), x - x_i) \rangle| =$$

$$|\langle (-1, \eta_0), (C(x) - C(x_0), x - x_0) \rangle + \langle (-1, \eta_0), (C(x_0) - C(x_i), x_0 - x_i) \rangle| \leq$$

$$\leq \frac{\epsilon}{4} \{|x - x_0| + |x_0 - x_i|\} \text{ for } x, x_i \in O_\delta.$$

On the other hand

$$C(x_i) - C(x) + \langle \eta_i, x - x_i \rangle = \langle (-1, \eta_i), (C(x) - C(x_i), x - x_i) \rangle \leq 0.$$

This implies that

$$\langle \eta_i - \eta_0, x - x_i \rangle =$$

$$\langle (-1, \eta_i), (C(x) - C(x_i), x - x_i) \rangle - \langle (-1, \eta_0), (C(x) - C(x_i), x - x_i) \rangle \leq$$

$$\leq \frac{\epsilon}{4} \{|x - x_0| + |x_0 - x_i|\}.$$

Choose x_i such that $|x_i - x_0| < \frac{\delta}{2}$.

Choose x such that $|x_i - x| > \frac{\delta}{2}$ and $x - x_i = \lambda(\eta_i - \eta_0)$ for some positive scalar λ. This implies that

$$\frac{1}{2} \epsilon \delta \leq |\eta_i - \eta_0||x - x_i| = \langle \eta_i - \eta_0, x - x_i \rangle \leq \frac{3}{8} \epsilon \delta - \text{ a contradiction.}$$

STEP 3. Choose $x \in AT(x_1; t_1, t_2)$. Consider a solution $(y(t), \eta(t))$ of the system $(8.2.1) - (8.2.2)$ such that

$$y(t_1) = x_1, \; y(t_2) = x, \; \eta(t_1) \in V(t_1, t_2).$$

Define the mapping $\varphi \; : \; AT(x_1; t_1, t_2) \mapsto V(t_1, t_2)$ as follows:

$$\varphi(x) = \eta(t_1).$$

Due to Lemma 1.4.1 $\varphi(x)$ is well defined. Consider a sequence $\{x_i\}$ such that x_i tends to x as $i \to \infty$. This implies that $\eta_i(t_2) \to \eta(t_2)$, hence $\varphi(x_i) \to \varphi(x)$. (Here $(y_i(t), \eta_i(t))$ is a solution of $(8.2.1) - (8.2.2)$ such that

$$y_i(t_1) = x_1, \ y_i(t_2) = x_i, \ \eta_i(t_1) \in V(t_1, t_2). \ )$$

This yields the continuity of $\varphi(x)$. Consider two different points $x \neq x'$ which belong to $AT(x_1; t_1, t_2)$. Corresponding solutions $(y(t), \eta(t))$ and $(y'(t), \eta'(t))$ of the system $(8.2.1)$, $(8.2.2)$ with $y(t_1) = y'(t_1)$ must be different at t_1. This implies that $\eta(t_1) \neq \eta'(t_1)$, in other words $\varphi(x) \neq \varphi(x')$. This completes the proof.

Lemma 8.3.2 For each $x_1 \in R^n$ the set $V(t_1, t_2) - \mathcal{N}(x_1)$ is an open subset of $V(t_1, t_2)$.

Proof. In order to prove the assertion we wish to define a continuous mapping

$$\psi : V(t_1, t_2) \mapsto R \times AT(x_1; t_1, t_2).$$

Let η be an element of $V(t_1, t_2)$. Denote by $(y(t), \xi(t))$ a solution $(y(t; x_1, \eta), \xi(t; x_1, \eta))$ of the system of differentiable equations $(8.2.1)$, $(8.2.2)$ with $y(t_1) = x_1$ and $\xi(t_1) = \eta$. Denote the corresponding control function $v(t, y(t), \eta(t))$ (see Definition 8.2.1) by $w(t)$. Finally, define

$$\psi(\eta) = (\int_{t_1}^{t_2} f(t, y(t), w(t)) dt, y(t_2)).$$

For each $\eta \in V(t_1, t_2) - \mathcal{N}(x_1)$ a trajectory $y(t; x_1, \eta)$ is not an optimal trajectory of $(8.1.1)$, $(8.1.2)$. This implies, that $\psi(\eta)$ belongs to the interior of the extended attainable set. This means, that $V(t_1, t_2) - \mathcal{N}(x_1)$ is a preimage of the open set. This finishes the proof.

The connectedness of $V(t_1, t_2)$ implies the following:

Corollary 8.3.2 The relation $\mathcal{N}(x_1) = V(t_1, t_2)$ holds for each $x_1 \in R^n$.

We now state and prove the main result of the Appendix:

Theorem 8.3.1 Each solution of $(8.2.1)$, $(8.2.2)$ generates an optimal trajectory of $(8.1.1)$, $(8.1.2)$. In other words: In the case of the linear control system with the convex cost the set of optimal trajectories coincides with the set of extremal trajectories.
Proof. An easy consequence of Corollary 8.3.2.

We call the system of differential equations $(8.2.1) - (8.2.2)$ the optimal equivalent of the control system $(8.1.1) - (8.1.2)$.

References

[1] L. D. Berkovitz, Optimal Control Theory. Springer-Verlag, 1974.

[2] F. H. Clarke, Optimization and Nonsmooth Analysis. John Wiley, 1983.

[3] E. A. Coddington, N. Levinson, Theory of Ordinary Differential Equations. McGraw-Hill, Inc., 1955.

[4] V.F. Dem'yanov, The solutions of some optimal control problems. SIAM Journal on Control, vol. 6, 1968, pp. 59-72.

[5] H. Federer, Geometric Measure Theory. Springer-Verlag, 1979.

[6] I.M. Gelfand, S.V. Fomin, Calculus of Variations. Prentice–Hall, Inc., 1963.

[7] J.-L. Goffin, A. Haurie, Pareto optimality with nondifferentiable cost functions. Conference on Multiple Criteria Decision Making. Jouy-en-Josas, France, 1975. Lecture notes in economics and mathematical systems edited by H. Thiriez and S. Zionts, Springer-Verlag, 1976, pp. 232-245.

[8] Ha Huy Vui, Sur les points d'optimum de Pareto local á determination finie ou infinie. C.R. Acad. Sc. Paris t.290 (21 Avril 1980) Serie A–685, pp. 685-688.

[9] A.D. Ioffe, V.M. Tihomirov, Theory of Extremal Problems. North–Holland, 1979.

[10] R.E. Kalman, The Theory of Optimal Control and the Calculus of Variations. Part 16 from Mathematical Optimization Techniques edited by Richard Bellman. University of California Press, Berkeley and Los Angeles, 1963.

[11] E.B. Lee, L. Markus, Foundations of Optimal Control Theory. John Wiley, 1967.

[12] N.G. Lloyd, Degree Theory. Cambridge University Press, 1978.

[13] R.R. Mohler, Bilinear Control Processes. Academic Press, 1973.

[14] C. Olech, Extremal solutions of control system. Journal of Differential Equations, vol. 2, 1966, pp. 74-101.

[15] L.S. Pontryagin, V.G. Boltyanskii, R.M. Gamkelidze, E.F. Mishchenko, The Mathematical Theory of Optimal Processes. John Wiley, 1962.

[16] R. T. Rockafellar, Convex Analysis. Princeton University Press, Princeton, New Jersey 1970.

[17] W. Rudin, Real and Complex Analysis. McGraw–Hill, Inc., 1974.

[18] W. Rudin, Functional Analysis. McGraw–Hill, Inc., 1973.

[19] S. Smale, Sufficient condition for an optimum. Dynamical systems–Warwick 1974 (Proc. Sympos. Appl. Topology and Dynamical Systems, Univ. Warwick, Coventry, 1973/1974 presented to E.C. Zeeman on his fiftieth birthday), pp. 287-292. Lecture Notes in Math. Vol. 468, Springer, Berlin, 1975.

[20] S. Smale, Optimizing Several Functions, Manifolds-Tokyo (Proc. Internat. Conf. on Manifolds and Related Topics in Topology), pp. 69-75, Univ. Tokyo Press, Tokyo, 1975.

[21] M. Spivak, Calculus on Manifolds. W. A. Benjamin, Inc., 1965.

[22] V. Zeidan, Extended Jacobi sufficiency criterion for optimal control. SIAM Journal on Control, vol. 22, 1984, pp. 294-301.

INDEX